Thermal Use of Shallow Groundwater

Thermal Use of Shallow Groundwater

Fritz Stauffer, Peter Bayer, Philipp Blum,
Nelson Molina-Giraldo, and Wolfgang Kinzelbach

CRC Press
Taylor & Francis Group
Boca Raton London New York

CRC Press is an imprint of the
Taylor & Francis Group, an **informa** business

CRC Press
Taylor & Francis Group
6000 Broken Sound Parkway NW, Suite 300
Boca Raton, FL 33487-2742

First issued in paperback 2017

© 2014 by Taylor & Francis Group, LLC
CRC Press is an imprint of Taylor & Francis Group, an Informa business

No claim to original U.S. Government works

ISBN-13: 978-1-4665-6019-2 (hbk)
ISBN-13: 978-1-138-07785-0 (pbk)

Visit the Taylor & Francis Web site at
http://www.taylorandfrancis.com

and the CRC Press Web site at
http://www.crcpress.com

Contents

Preface   *xi*
Acknowledgments   *xiii*
Authors   *xv*
Symbols   *xix*

1 Introduction   1

   1.1   Motivation for the thermal use of
        underground or groundwater systems 2
   1.2   Importance of the local conditions 4
      1.2.1   Thermal regime 4
      1.2.2   Hydrological and hydrogeological conditions 9
   1.3   Technical systems 10
      1.3.1   Heat pumps 10
      1.3.2   Closed- and open-loop systems 12
   1.4   Energy demand and energy production 16
   1.5   Management of underground resources 19
      1.5.1   Seasonal operation of technical installations 19
      1.5.2   Water supply and thermal use 20
   1.6   Impact on groundwater quality and ecology 20
   1.7   Geotechnical issues 21
   1.8   Regulatory issues 24
      1.8.1   Swiss regulation 25
      1.8.2   Austrian regulation 29
      1.8.3   British regulation 29
      1.8.4   German regulation 30
   1.9   Challenges related to design and management 31
   1.10 Scope of the book 32
   References 32

2 Fundamentals   37

2.1   *Theory of water flow and heat transport in the subsurface 37*
    2.1.1   *Modeling hydraulic processes in porous media 37*
        2.1.1.1   *Flow in saturated and unsaturated*
                porous media, Darcy's law 37
        2.1.1.2   *Water mass balance, volume*
                balance, flow equation 44
        2.1.1.3   *Initial and boundary conditions 49*
        2.1.1.4   *Two-dimensional flow models for*
                saturated regional water flow 50
    2.1.2   *Modeling thermal processes in porous media 52*
        2.1.2.1   *Heat storage, heat capacity,*
                and advective heat transport 53
        2.1.2.2   *Heat conduction 54*
        2.1.2.3   *Dispersive and macrodispersive*
                heat transport 61
        2.1.2.4   *Heat transport equation 68*
        2.1.2.5   *Initial and boundary conditions 72*
        2.1.2.6   *Concepts for BHEs 75*
        2.1.2.7   *Coupling thermal transport*
                with hydraulic models 79
        2.1.2.8   *Two-dimensional heat transport models 80*
    2.1.3   *Integral water and energy balance*
        equations for aquifers 81
        2.1.3.1   *Rough estimation of the*
                potential of an unconfined
                shallow aquifer for thermal use 85
2.2   *Thermal property values 87*
    2.2.1   *Heat capacity and thermal conductivity values 87*
References 93

3 Analytical solutions   101

3.1   *Closed systems 105*
    3.1.1   *Instantaneous point source—three-*
        dimensional conduction 105
    3.1.2   *Moving point source—three-dimensional*
        conduction and advection 105

3.1.3   ILS—two-dimensional conduction 106

3.1.4   Infinite cylindrical source—
two-dimensional conduction 109

3.1.5   FLS—three-dimensional conduction 114

3.1.6   Finite cylindrical source—three-
dimensional conduction 118

3.1.7   Moving ILS—two-dimensional
conduction and advection 119

3.1.8   Moving FLS—three-dimensional
conduction and advection 126

3.1.9   Infinite plane source—one-
dimensional conduction 130

3.1.10   Moving infinite plane source—one-
dimensional conduction and advection 132

3.1.11   Steady-state injection into an aquifer
with thermally leaky top layer 134

3.1.12   Harmonic temperature boundary
condition for one-dimensional
conductive–advective heat transport 135

3.1.12.1   One-dimensional vertical
conductive heat transport 135

3.1.12.2   One-dimensional horizontal
conductive/dispersive–
advective transport 136

3.1.12.3   Horizontal layer embedded in
conductive bottom and top layer 138

3.2   Open systems 140

3.2.1   Analytical solution for steady-state
flow in multiple well systems 142

3.2.1.1   Double well system in
uniform flow field 145

3.2.2   Linear flow 152

3.2.3   Radial flow, infinite disk source 155

3.2.4   Natural background groundwater flow 156

References 158

4 Numerical solutions   163

4.1   Two-dimensional horizontal numerical solutions 167

4.1.1   Analogy with solute transport models 170

4.1.2   *Analysis of steady-state open system in rectangular aquifer 171*
4.1.2.1   *Scaled solution for open system in rectangular aquifer 173*
4.2   *Multidimensional numerical solutions 175*
4.2.1   *Principles of the finite difference method for heat transport 176*
4.2.2   *Principles of the finite element method for heat transport 181*
4.2.3   *Principles of the finite volume method for heat transport 183*
4.2.4   *Principles of the method of characteristics for heat transport 183*
4.2.5   *Principles of the random walk method for heat transport 184*
4.3   *Strategy for coupled flow and heat transport 185*
4.4   *Some available codes for thermal transport modeling in groundwater 186*
References 190

5   **Long-term operability and sustainability**   197

5.1   *Systems in low permeable media 197*
5.2   *Thermal evolution in aquifers 202*
5.3   *Further criteria of sustainability 204*
References 207

6   **Field methods**   209

6.1   *Hydrogeological field methods 209*
6.2   *Thermal response tests 210*
6.2.1   *Development of TRTs 210*
6.2.2   *Setup and application of TRTs 212*
6.2.3   *Evaluation of TRTs 213*
6.2.3.1   *Analytical models 214*
6.2.3.2   *Numerical models 218*
6.3   *Thermal tracer test 220*
References 224

7 Case studies   229

7.1   *Case study Altach (Austria) 232*
7.2   *Limmat Valley aquifer Zurich (Switzerland) 239*
7.3   *Bad Wurzach (Germany) 243*
References 248

Index   *251*

Preface

The thermal use of the shallow subsurface is increasingly promoted and implemented as one of many promising measures for saving energy. The energy extracted from such systems is referred to as shallow geothermal energy or low-enthalpy energy. Open and closed systems are distinguished usually consisting of boreholes combined with heat pumps. A series of questions arises with respect to the design, the management of underground and groundwater heat extraction systems, such as the sharing of the thermal resource, the long-term sustainability of the thermal use, and the assessment of its long-term potential. For the proper design of thermal systems, it is necessary to assess their impact on underground and groundwater temperatures.

The theoretical basis of heat transport in soil and groundwater systems is therefore introduced, and the essential thermal properties are discussed. In the planning and design of geothermal systems, hydrogeological and thermal site investigations have to be combined with modeling. Therefore, a series of mathematical tools and simulation models based on analytical and numerical solutions of the heat transport equation are presented. Finally, some case studies are introduced for illustration.

The book is directed toward MSc students in civil or environmental engineering, engineering geology, and hydrogeology and junior professionals. It provides a platform of principles and outlines the essential models and parameters to assess and design technical systems for the thermal use of the shallow underground.

MATLAB® is a registered trademark of The MathWorks, Inc. For product information, please contact:

The MathWorks, Inc.
3 Apple Hill Drive
Natick, MA 01760-2098 USA
Tel: 508-647-7000
Fax: 508-647-7001
E-mail: info@mathworks.com
Web: www.mathworks.com

Acknowledgments

We would like to thank everybody who supported us in our book project. In particular, we acknowledge the research of former members of the Hydrogeothermics Research Group: Gita Brandstetter, Stefanie Hähnlein, Joszef Hecht-Méndez, Karl Urban, Selçuk Erol, and Ke Zhu. Furthermore, we are grateful to Gabriele Moser, Gianni Pedrazzini, and Valentin Wagner, who have supported us in the editing of text and figures. Special thanks go to Signhild Gehlin and Markus Kübert for providing the photos and data of the thermal response tests. For the support, data, and excellent cooperation at the geothermal test site in Bad Wurzach, we would like to thank Gerhard Bisch, Jürgen Braun, Nobert Klaas, and Christoph Knepel. Moreover, we acknowledge the excellent work of the master students Niculin Cathomen, Esther Müller, and Daniel Ott in environmental engineering, which was utilized for the presentation of the regional case studies.

Authors

Fritz Stauffer, Prof. Dr., is a retired professor at the Institute of Environmental Engineering at ETH Zurich, Switzerland. He was born in 1947 and is a citizen of Switzerland. He was a senior research officer and lecturer in groundwater and hydromechanics. He studied rural engineering at ETH Zurich, where he obtained his diploma in 1971 and his doctorate in 1977. The PhD thesis—under the supervision of Professor Dr. T. Dracos—was in the field of unsaturated flow in porous media. From 1981 to 2010, Fritz Stauffer taught courses in groundwater and organized a series of 20 international scientific short courses in groundwater management as well as several international symposia at ETH Zurich. In 1996, he was a visiting scholar at Stanford University. In 2001, he was awarded the title of professor by ETH Zurich. His research interests are in flow and contaminant transport in groundwater including the capillary zone.

Apart from using experimental techniques, he mainly focuses on mathematical modeling as well as geostatistics and stochastic modeling of flow and transport processes in highly heterogeneous aquifers. Main applications are the quantification of the uncertainty in the localization of groundwater protection zones, the interaction between rivers and aquifers, thermal processes in groundwater, the coupled flow and heat transport in porous media with phase change (as in the thawing permafrost of rock glaciers), the thermal use of groundwater, and two-phase flow processes in porous media. He retired from ETH in January 2012.

Peter Bayer, Dr., is a senior research associate at the Department of Earth Sciences at ETH Zurich, Switzerland. He was born in 1972 and is a German citizen. He graduated in 1999 from the Center for Applied Geosciences of the University of Tübingen (Germany) and earned his PhD from the same institution in 2003. From 2008 to 2010, Peter Bayer was a EU Intra-European Marie Curie Fellow at ETH Zurich hosted by the Institute of Environmental Engineering. His work was aimed at the development of algorithmic optimization procedures and their implementation to solve problems related to applied hydrogeology and geothermics.

Among his main scientific contributions in the area of thermal use of shallow aquifers are the development and application of life cycle–based concepts for analyzing environmental aspects of low-enthalpy geothermal systems. Together with coauthors P. Blum and N. Molina-Giraldo, he investigated the influence of groundwater flow on borehole heat exchangers by analytical and numerical simulation as well as through field studies. This work was accompanied by enhancements to thermal field investigation techniques in shallow aquifers, in particular, the application and interpretation of thermal response tests under conditions with significant groundwater flow velocity. Peter Bayer has been working on thermal processes in aquifers on the lab scale, the several meters field scale, as well as the large scale of urban subsurface temperature evolution. His research has been published in more than 75 scientific contributions, 50 of which are listed in the Web of Science.

Philipp Blum, Jun.-Prof. Dr., born in Ulm in 1972, is currently an assistant professor (junior professor) for engineering geology at the Karlsruhe Institute of Technology (KIT). From 1993 to 1996, he studied geology at the University of Heidelberg. In 1997, as part of the Erasmus programme, he joined the School of Earth Sciences at Cardiff University in Wales. From 1996 to 2000, he continued his studies in applied geology at the University of Karlsruhe, where he received his diploma in 2000. In 2003, as part of the international research project DECOVALEX, he received his PhD on hydromechanical processes in fractured rock at the School of Earth Sciences at the University of Birmingham (UK). From 2003 to 2005, he was working for URS Germany as a project manager and hydrogeologist. From 2006 to 2010, he was an assistant professor for hydrogeothermics at the University of Tübingen (Germany), where in 2010, he received his habilitation in applied geology on thermohydromechanical and chemical (THMC) processes in porous and fractured aquifers. He published more than 40 peer-reviewed publications, of which 36 are also listed in the Web of Science. His current research interests focus on contaminant hydrogeology, shallow geothermal energy, and engineering geology in porous and fractured rocks.

Nelson Molina-Giraldo, Dr., born in 1981 in Colombia, is a groundwater modeler at Matrix Solutions, Inc., Canada. He obtained his first degree in environmental engineering at the University of Antioquia, and then he moved to Germany and completed a master programme in applied environmental geosciences (AEG) at the University of Tübingen, Germany. He received his

PhD at the same university in 2011, where he conducted research into heat transport modeling in shallow aquifers. Currently, he has been working on analytical and numerical modeling to assess the feasibility of groundwater withdrawal forecasts for operational management and regulatory needs. He has been also implementing groundwater–surface water monitoring programs to attempt to measure changes in groundwater–surface water interaction based on temperature measurements.

Wolfgang Kinzelbach, Prof. Dr., born in 1949 in Germany, is a full professor of hydromechanics and groundwater at ETH Zurich, Switzerland. He obtained his first degree in physics at Munich University and then turned to environmental engineering in Stanford, California. He earned his PhD at Karlsruhe University in 1978, with a thesis on managing waste heat emissions by power stations on the Rhine River. After professorships at Kassel University and Heidelberg University, he joined ETH in 1996. He has been working on groundwater themes including the modeling of pollutant transport, environmental tracers, design of remediation measures, and management of water resources for more than 20 years. His more recent work focuses on real-time modeling and control of well fields and sustainable management of water resources in arid environments. He is the author or a coauthor of more than 200 publications listed in the Web of Science. He was awarded the Henry Darcy Medal of the European Geophysical Union, the Prince Sultan International Prize for Water, and the Muelheim Water Award. He is a fellow of the American Geophysical Union.

Symbols

Bold symbols represent vectors and tensors, whereas italic symbols represent scalar quantities and variables.

$_a$:   Subscript a: air
A: Area (m^2)
$_b$:   Subscript b: borehole
B: Boundary
c: Solute concentration in water (kg m^{-3})
C_m: Volumetric heat capacity of porous medium or aquifer (J m^{-3} K^{-1}) or (W s m^{-3} K^{-1})
c_s: Specific heat capacity or specific thermal capacity, of solid material (J kg^{-1} K^{-1})
C_s: Volumetric heat capacity of solid material (J m^{-3} K^{-1}) or (W s m^{-3} K^{-1})
c_w: Specific heat capacity or specific thermal capacity of water (J kg^{-1} K^{-1}) or (W s m^{-3} K^{-1})
C_w: Volumetric heat capacity of water (J m^{-3} K^{-1})
D: Aquifer domain
$\mathbf{D_h}$:   Hydrodynamic dispersion tensor for solute transport (m^2 s^{-1})
$\mathbf{D_t}$:   Thermal diffusion tensor or thermal diffusivity tensor (m^2 s^{-1})
$D_{t,L}$:   Longitudinal thermal diffusion coefficient (m^2 s^{-1})
$D_{t,T}$:   Transversal thermal diffusion coefficient (m^2 s^{-1})
E: Energy (J) or (W s)
$_f$:   Subscript f: fluid
f: Depth to groundwater (m)
Fo: Fourier number (–)
g: Gravitational acceleration constant (scalar) (m s^{-2})
$\mathbf{g}$:   Gravitational acceleration gradient (m s^{-2})
H: Length of vertical borehole heat exchanger (m)
h_w: Piezometric head (m)
$_i$:   Subscript i: ice
I: Recirculation rate between two wells (m^3 s^{-1})
I_{hor}:   Horizontal flow gradient (–)

j: Specific heat flux (W m^{-2})
J: Heat flux (W) or (J s^{-1})
j_{disp}:   Dispersive (specific) heat flux (W m^{-2})
$\mathbf{k}$:   Permeability of aquifer (tensor) (m^2)
$\mathbf{K_w}$:   Hydraulic conductivity of aquifer (tensor) (m s^{-1})
L: Length scale (m)
L_f: Latent heat of melting/freezing (J kg^{-1}), 3.34 10^5 J kg^{-1} for water/ice
m: Aquifer thickness (m)
m_{VG}:   van Genuchten parameter (–)
$_{-n}$:   Subscript n: normal direction
$\mathbf{n}$:   Unit normal vector (m)
N: Recharge rate per unit surface area (m s^{-1})
n_{VG}:   van Genuchten parameter (–)
$_{-p}$:   Subscript p: pipe
p_b: Air entry pressure (Pa)
P_t: Heat production per unit volume (W m^{-3})
p_w: Water pressure (Pa)
$\mathbf{q}$:   Specific discharge vector, water discharge rate through unit area
         (gradient) (m^{3}m^{-2} s^{-1})
Q: Water discharge rate (m^3 s^{-1})
q_{tb}:   Heat flow rate per unit length of the borehole (= J/H) (W m^{-1})
R: Dimensionless cylindrical radius (–)
R_c: Solute retardation factor (–)
R_{t_ret}:   Thermal retardation factor (–)
R_{tb}:   Thermal borehole resistance (K W^{-1} m^{-1})
R_{tw}:   Thermal radius of influence (m)
R_w: Radius of influence of a well (m)
$_{-r}$:   Subscript r: residual
r: Radius (m)
r_b: Borehole radius (m)
r_p: Pipe radius (m)
$_{-s}$:   Subscript s: solid
s: Length (m)
S: Storativity of aquifer, specific yield of unconfined aquifer (–)
S_s: Specific storativity of aquifer (m^{-1})
S_w: Saturation degree of water (–)
$S_{w,r}$:   Residual saturation degree of water (–)
$_{-t}$:   Subscript t: thermal
t: Time (s)
T: Temperature (°C or K; 0°C = 273.15 K)
T_0:   Initial or undisturbed temperature (°C)
T_{inj}:   Injection temperature (°C)
$\mathbf{u}$:   Mean flow velocity (gradient) (m s^{-1})
$\mathbf{u_t}$:   Thermal velocity (gradient) (m s^{-1})

$_v$:	Subscript v: vapor
V:	Volume (m^3)
vf:	Volumetric fraction (–)
$_w$:	Subscript w: water
w:	Water source/sink term, water volume per unit aquifer volume and unit time (m^3 m^{-3}s^{-1})
W:	Source/sink term in two-dimensional flow equation (m^3 m^{-2} s^{-1})
x:	x-coordinate (m)
$\mathbf{x}$:	Location vector (m), with coordinates (x, y, z)
y:	y-coordinate (m)
z:	z-coordinate (m), positive upward
α:	Angle (rad or °)
α_L:	Longitudinal dispersivity for solute transport (m)
α_T:	Transversal dispersivity for solute transport (m)
α_{VG}:	Van Genuchten parameter (m)
β:	Angle (rad or °)
β_L:	Longitudinal thermal dispersivity of aquifer (m)
β_T:	Transversal thermal dispersivity of aquifer (m)
γ:	Euler's constant (–) = 0.5772…
Δ:	Finite increment
θ:	Dimensionless temperature (–)
θ_a:	Volumetric air content (m^{-3} m^{-3})
θ_i:	Volumetric ice content (m^{-3} m^{-3})
θ_w:	Volumetric water content (m^{-3} m^{-3})
λ_{BC}:	Pore distribution index in the model of Brooks and Corey (–)
λ_{decay}:	First-order decay coefficient for solute transport (s^{-1})
λ_{disp}:	Thermal dispersion tensor (W m^{-1} K^{-1})
λ_{eff}:	Effective thermal conductivity of subsurface (W m^{-1} K^{-1})
λ_m:	Thermal conductivity of porous medium or aquifer (W m^{-1} K^{-1})
λ_s:	Thermal conductivity of solid material (W m^{-1} K^{-1})
λ_w:	Thermal conductivity of water (W m^{-1} K^{-1})
λ_{vert}:	Thermal conductivity of the overburden (W m^{-1} K^{-1})
μ:	Dynamic viscosity (Pa s)
ρ:	Dimensionless radius (m)
ρ_a:	Density of air (kg m^{-3})
ρ_i:	Density of ice (kg m^{-3})
ρ_s:	Density of the solid phase of the aquifer (kg m^{-3})
ρ_{rel}:	Relative density (–)
ρ_w:	Density of water (kg m^{-3})
τ:	Period (s)
φ:	Flow potential (m^2 s^{-1})
φ_r:	Angular coordinate (polar angle) (–)
ϕ:	Porosity of aquifer, volumetric fraction of pores in aquifer (m^{-3} m^{-3} or –)

χ: Scaled pumping rate (–)

ψ: Stream function (m^2 s^{-1})

ω: Angular frequency (s^{-1})

∇:   Gradient operator, applied to scalar quantity f: $\nabla f = \left(\dfrac{\partial f}{\partial x}, \dfrac{\partial f}{\partial y}, \dfrac{\partial f}{\partial z} \right)$

$\nabla\cdot$:   Divergence operator, e.g., applied to vector $\mathbf{v}$: $\nabla \cdot \mathbf{v} = \dfrac{\partial v_x}{\partial x} + \dfrac{\partial v_y}{\partial y} + \dfrac{\partial v_z}{\partial z}$

Chapter 1

Introduction

The **thermal use of shallow subsurface systems** is increasingly discussed, promoted, and implemented as one of many promising measures to reduce fossil fuel use. The energy extracted from such systems is referred to as shallow geothermal energy or low-enthalpy energy. These systems may consist of groundwater abstraction by pumping wells, energy generation, or abstraction with the support of heat pumps such as groundwater heat pump (GWHP) systems, and the reinjection of the cooled or warmed water into the aquifer (open system, Figure 1.1b). On the other hand, thermal use may also include pumping of groundwater for cooling purposes and reinjection of warm water. Both systems may even be seasonally combined. A related topic is heat storage (Dinçer and Rosen 2011). Since temperature in undisturbed shallow aquifers is around annual mean air temperature at a given location outside the range of influence of infiltrating rivers or lakes, energy production by ground-source heat pumps and GWHPs is attractive also at low air temperatures in winter. Thermal use of shallow underground (saturated or unsaturated zone) can also be accomplished by low-enthalpy geothermal heat exchanger systems combined again with heat pumps (closed systems or ground source heat pump systems [GSHP], Figure 1.1a). With respect to the thermal management of underground and groundwater systems, a series of questions arise in this context. What is the tolerable temperature increase or reduction of groundwater? What is the long-term usable energy potential of an aquifer? What is the long-term sustainability of the thermal use of groundwater? What are management problems with respect to thermal use? What are the geotechnical risks related with the thermal use of the underground? What harm comes from the thermal use of groundwater? Is groundwater quality and/or groundwater ecology affected? Is there a competition between drinking water production and thermal use? How much does a city heat up shallow groundwater? How can the temperature development be assessed? How is the heat balance affected? In order to answer these questions and to design thermal systems, it is necessary to provide methods to compute their effects on the development of temperature in the

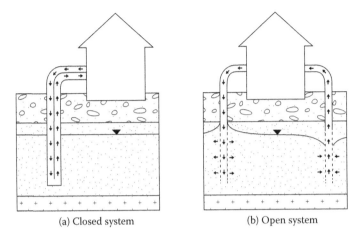

(a) Closed system   (b) Open system

Figure 1.1 (a) Scheme of closed-loop (GSHP system) and (b) open-loop (GWHP system) shallow geothermal systems.

underground and, in particular, the groundwater. Accordingly, the theoretical fundamentals of heat transport in groundwater systems are recalled, and the essential thermal properties and parameters are reviewed and discussed. Hydrogeological–thermal investigations have to be combined with modeling. Therefore, a series of mathematical tools and simulation models based on analytical and numerical solutions are presented and discussed. Case studies are shown for locations in Austria, Germany, and Switzerland. They concern the urban thermal energy use as well as heat storage and cooling.

1.1 MOTIVATION FOR THE THERMAL USE OF UNDERGROUND OR GROUNDWATER SYSTEMS

The prime motivation for the thermal use of shallow underground systems has clearly been the partial substitution of fossil fuel energy by increasing the overall efficiency of thermal power plants. It was seen as an efficient way to transform electricity from nuclear power stations into energy for space heating and thus save a multiple of that amount in fossil fuel. One of the main drivers is, no doubt, the price for energy production. An example is the oil crises between 1973 and about 1982, which led to increased energy prices in this period and triggered, among others, various activities toward thermal use of underground space and groundwater worldwide (e.g., Balke 1974, 1977). The expected long-term exhaustion of fuel reservoirs has been another concern in this context. Therefore, energy safety and independence are important motivations. Recent developments are mainly motivated by the global warming and the debate and discussion about a reduction of the carbon dioxide (greenhouse gases) concentration of the atmosphere (e.g., Blum et al. 2010; Bayer et al. 2012). For example, in 2008, the use of around 879,000 ground-source heat pump (GSHP) systems in nineteen European countries saved 3.7×10^6 t CO_2 (eq.) in comparison to conventional heating practice (Bayer et al. 2012). Hence, the thermal use of underground systems has been considered and discussed as an alternative source of energy.

The thermal use of shallow underground systems comprises the **utilization of the underground** (rock, dry, or unsaturated as well as saturated soil) or **pumped groundwater as heat source** (for heating purposes) **or heat sink** (for cooling purposes). In principle, it is possible to directly use soil air or groundwater together with heat exchangers for heating or cooling depending on the prevailing temperatures. However, **thermal systems combining heat exchangers and heat pumps** are in the focus of current activities.

The extraction of heating or cooling energy or the extraction of groundwater and reinfiltration of cooled or warmed water produces **thermal anomalies**, which propagate in the subsurface environment. One aspect consists of **thermal exploitation** of the resource, which represents thermal mining. In this context, the possible **thermal overexploitation** is of concern, that is, excessive cooling or even freezing and warming of the underground. The other aspect consists of the **induction of thermal fluxes**, which are caused by the thermal anomaly. Main fluxes in shallow groundwater systems are thermal fluxes from the soil surface to the aquifer or from the infiltration of rainwater through the surface and the infiltration of surface water as well as the geothermal flux. In the long run, a **stable thermal yield** can be expected, however, at the cost of long-term shift (increase or decrease) of the temperature in the soil and groundwater system. Care has to be taken that this equilibrium is not reached at a temperature at the extraction point, which is so low that the heat pump system ceases to function.

A well-known phenomenon of urban environments above shallow groundwater systems is the so-called **urban heat island effect** (e.g., Landsberg 1956; Balke 1974; Ferguson and Woodbury 2007; Zhu et al. 2010). As a result of land use change and urban infrastructure, air temperature has increased, typically by 2 to 5 K or even more, compared to rural areas (e.g., Oke 1973). In Ankara, for example, mean soil temperatures in 5–50 cm depths were found to be 1.8–2.1 K higher in the city area than at rural stations (Turkoglu 2010). As a consequence, the temperatures of soil surface, soil, and groundwater have been raised as well (e.g., Balke 1974; Changnon 1999; Ferguson and Woodbury 2007; Taniguchi et al. 2007). Moreover, constructions like **buildings** (including heated basements) or **sewer systems** can heat underground and groundwater significantly, mainly due to heat conduction. All these effects lead to thermal anomalies. Menberg et al. (2013) introduced an indicator, the 10%–90% quantile range of the urban heat island intensity ($UHII_{10-90}$), to quantify and compare different UHI intensities in the groundwater. For the six German cities investigated, the $UHII_{10-90}$ ranged between 1.9 and 2.4 K. The **effect of urbanization on shallow groundwater temperatures** was investigated by Taylor and Stefan (2009). Their analysis showed that groundwater temperature in fully urbanized regions is up to 3 K higher than in agricultural areas. According to Taylor and Stefan (2009), **pavements** (e.g., asphalt strips) are a main cause for the excess temperature. Ferguson and Woodbury (2004) conjectured, for an urban field site in Canada, that temperature changes could be largely attributed to **heat loss from buildings**.

The **heating effect of buildings** on the subsurface depends on a series of factors. It depends on the structure of the building, the foundation, the depth of the foundation, the existence of cellars (basements), the number of underground levels, the building materials, the existence of insulation of walls and floor plates or ceilings, the efficiency of the insulation, the heating of the underground floors (by heating elements or by ventilation), etc. Of course, the physical properties of the underground material and the depth to groundwater are of importance as well. Even in unheated basements, the annual mean temperature is typically higher than the annual mean soil surface temperature.

Within the concept of thermal use of underground systems, this **surplus energy offers an additional potential for energy abstraction**. Allen et al. (2003) emphasized that using a hydrogeothermal source for space heating has high development potential in urban heat islands with high yielding aquifers. They estimated that a well that yields 20 L s^{-1} of 13°C groundwater from an Irish urban aquifer can generate 856 kW of heat, which can satisfy the heating demand of a 12,000 m^2 building with a peak heating intensity of 70 W m^{-2}. On the other hand, the phenomenon may limit possibilities for seasonal heat storage and the use of groundwater for cooling. In the city of Frankfurt, for example, where cooling is of primary interest, the elevated groundwater temperatures (by several degrees, locally up to >30°C in the down-gradient vicinity of a coal-burning power plant) are unfavorable for cooling purposes (Menberg et al. 2013).

1.2 IMPORTANCE OF THE LOCAL CONDITIONS

1.2.1 Thermal regime

A principal consideration for the use of shallow geothermal or low-enthalpy energy, besides the identification of the various physical processes involved, is the **thermal regime**. As shallow underground, we define the first few decameters below ground surface. Such systems are highly influenced by the local meteorological conditions. Approximately, the temperature of shallow underground systems at a considered location is close to the mean annual soil temperature. Key information is the vertical above and below ground temperature profile (Figure 1.2). The annual mean air temperature at a location is usually measured at an elevation of 1–2 m above ground surface. It ranges from about 0°C in polar zones to about 24°C in tropical regions. In the temperate climate zone, it is around 8°C to 10°C. This reference air temperature is typically lower than the mean annual ground surface temperature (GST). With increasing depth, the temperature normally increases due to the local geothermal heat flux, which typically ranges between 0.05 and 0.11 W m^{-2} depending on the local geological setting (Pollack et al. 1993). The related vertical temperature gradient is around 0.03 K m^{-1}. Moreover, cryologists observed a thermal offset within the active zone of frozen ground close to soil surface (Williams and Smith 1989). In general, soil temperature depends mainly on latitude, elevation above sea level, exposition of the site, groundwater flow, and the local geothermal heat flow. Seasonal temperature fluctuations are usually perceptible only to a depth of about 10 m below ground surface. Below this depth, the thermal regime is highly damped or even stable. An example of the thermal profile measured close to soil surface at a meteorological station in

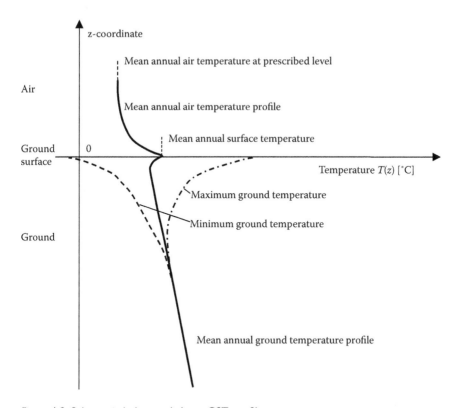

Figure 1.2 Schematic below and above GST profile.

Zurich-Affoltern (Switzerland) is presented in Figure 1.3. It shows that the **annual mean temperature at soil surface is about 2 K** higher than air temperature in this case, with a long-term mean air temperature of 8.5°C and a long-term mean soil surface temperature of about 10.5°C. The monthly average temperature of air (2 m above ground level) and soil (5 cm below ground level) at the meteorological station Zurich-Affoltern (Switzerland) over the period 2000–2010 is depicted in Figure 1.4. Accordingly, a temperature difference of 2 K persists over the year except for the months March and April at this station (mean difference of 1.5 K over the period of 2000–2010).

A similar **difference between mean air and soil surface temperature** was found by Wu and Nofziger (1999) in their investigations of bare soil in northern China. The three-year annual mean surface soil temperature was about 2.0 K higher than that of the air temperature. Similar results for bare soils were already reported by Fluker (1958). While the average annual air temperature was 20.8°C, it was 24.1°C at a soil depth of 5 cm. Moreover, the corresponding maximum average temperatures were 30.0°C and 35.2°C, respectively, and the minimum temperatures were 10.5°C and 11.1°C, respectively. This observation implies corresponding temperature amplitudes of 9.8 and 12.1 K. Measurements of the soil temperature regime in the United States (USDA 1999) showed a typical difference of 1 K between annual air and soil surface temperature. For 14 stations within the United States, this difference ranged between 0.7 and 2.9 K. Putnam and Chapman (1996) inspected a one-year record of ground surface and air temperature measurements in northwest Utah (United States). Mean GSTs were seen to be higher for bare granite (11.3°C) than for partially shaded neolith (9.5°C), and both were above the mean annual air temperature at a height

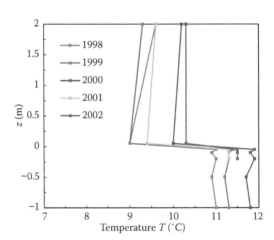

Figure 1.3 Above and below ground temperature profiles at the meteorological station Zurich-Affoltern (Switzerland); data MeteoSwiss.

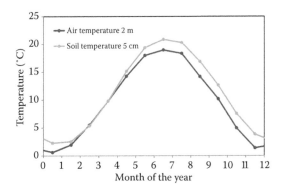

Figure 1.4 Monthly average temperature of air (2 m above ground surface) and soil (in 5 cm depth) at the meteorological station Zurich-Affoltern (Switzerland); period 2000–2010; data MeteoSwiss.

of 2 m above ground (8.8°C). For a soil covered by crops, the temperatures in the profile are expected to be lower than those under a bare surface due to shading and evaporation. Influence of precipitation and snow cover is negligible in the arid environment of this field site.

Lee (2006) found for **urban environments** (City of Seoul, South Korea) that mean annual ground temperatures (13.7°C–13.8°C) were slightly higher than the mean air temperature (12.9°C).

Based on hourly **satellite-derived land surface temperature** (LST) and measured air temperature (T_a) data at 14 stations of the US Climate Network, Gallo et al. (2011) investigated the relationship between air and LST. Vegetation at each station is specified as grass or low vegetation ground cover. LST was greater than T_a for both the clear and cloudy conditions; however, the differences between LST and T_a were significantly less for the cloudy-sky conditions. Mean differences of less than 2 K were observed under cloudy conditions for the stations, as compared with a minimum difference of greater than 2 K (and as great as 7 K) for the clear-sky conditions. The results suggest that the relationship between LST and T_a, even under cloudy conditions, can vary with location. There is a cyclic increase and decrease in the clear-sky hourly difference between LST and T_a, which generally follows the daily and seasonal cycle of solar radiation. Under cloudy conditions, the difference is fairly stable throughout the day. Smerdon et al. (2006) concluded that the main differences between GST and surface air temperature are caused by summer evapotranspiration and winter cryogenic effects (snow, ice).

The **temperature at soil surface** of a particular site, that is, just within the soil, mainly depends on the net radiation flux and the heat transfer fluxes between soil surface and the air layer above it, as well as energy fluxes into or from the ground. Net radiation is commonly expressed by the **heat balance at the soil surface** (Figure 1.5) as follows (e.g., Williams and Smith 1989; Kollet et al. 2009):

$$J_{net\_radiation} = J_{sensible\_heat} + J_{latent\_heat} + J_{ground\_heat} \qquad (1.1)$$

Note that Equation 1.1 holds for both short- as well as long-term evaluations. The **net radiation** ($J_{net\_radiation}$) is the incoming shortwave and longwave (IR) radiation minus the outgoing longwave radiation. **Soil surface temperature** is also influenced by the heat transfer between soil surface and air layer, which is composed of the sensible heat flux ($J_{sensible\_heat}$) due to the movement of air, the latent heat flux (J_{latent_heat}) due to evapotranspiration and condensation, as well as the ground heat flux (J_{ground_heat}). Energy fluxes from the surface into the ground (referred to as ground heat flux including negative fluxes from the ground to the surface) comprise heat conduction in solids, soil air and soil water, advective heat flux in soil air (including vapor), and flowing soil water (from rainwater). **Ground heat flux** shows a distinct diurnal and seasonal variability with inflowing and outflowing components, while in the long run, the average ground heat flux is basically identical to the geothermal heat flux.

The thermal use of shallow underground systems will certainly also play a role in the energy balance at soil surface by affecting the soil surface temperature. Taking the resulting **seasonal soil surface temperature**, based on measurement, as reference and thermal boundary condition leads to a pragmatic, simplified formulation of the complex situation.

The unperturbed **temperature of shallow groundwater systems,** with depth to groundwater as well as saturated thickness of up to tens of meters, is often close to the mean annual soil surface temperature. The influence of the geothermal heat flux is usually weak in shallow systems. However, **the thermal regime of shallow groundwater systems can be,** more or less,

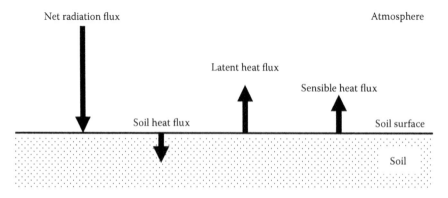

Figure 1.5 Energy fluxes toward soil surface.

strongly influenced by infiltrating surface water from rivers, streams, and lakes. This influence manifests itself by additional temperature fluctuations. Sufficiently far away from infiltrating surface water (order of a few 100 m horizontally), the fluctuations vanish. Moreover, groundwater systems very close to the soil surface exhibit an influence of the seasonal temperature fluctuation at the soil surface.

The fact that temperatures close to the mean annual soil temperature, at a depth of about 10 m or more, are relatively stable provides an important prerequisite for the thermal use of underground systems by heat exchanger systems with the help of heat pumps. This concerns both heating and cooling systems for air conditioning and space heating (Figure 1.6a) as well as cooling. Moreover, direct heat injection from buildings (geo-cooling) without heat pump (Figure 1.6b) is also possible.

1.2.2 Hydrological and hydrogeological conditions

In principle, any underground space may be used thermally. However, the choice of a specific technical system, that is, the installation of borehole heat exchangers (BHEs) or pumping and reinjection of groundwater with heat extraction, strongly depends on the local hydrological and hydrogeological conditions.

For closed systems, low hydraulic conductivity formations are preferred. The main reason is to prevent groundwater pollution caused or enabled by leakage of working fluids of the installation. In this context, highly conductive karst and fractured rock formations are less suited. A prominent example for inappropriate installation due to another reason is the closed system planned

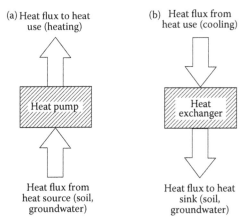

Figure 1.6 Schematic heat flux for (a) space heating with heat pump; (b) space cooling with heat exchanger only.

for the town of Staufen in the Upper Rhine Graben in southwest Germany. Down to a depth of 140 m, seven boreholes were drilled in 2007 into gypsum and anhydrite-bearing Keuper formations (Goldscheider and Bechtel 2009). The boreholes established **hydraulic contact** between confined groundwater and anhydride, and the consequence was substantial volume increase from gypsum swelling. Within a couple of weeks, **uplift of the ground** in the range of 1 cm month^{-1} caused severe damage to the buildings of the town.

Pumping and reinjection of groundwater in **open systems requires that the hydraulic conductivity of the aquifer is sufficiently large** in order to be efficient. Further important parameters are the geologic structure, the porosity, the water content, and the mineral composition of the aquifer material. Important hydrological conditions are the local flow direction, the recharge conditions, the location of surface water bodies, the depth to groundwater, and the fluctuation of the groundwater table. Hydrothermal conditions comprise the soil and water temperature, their fluctuation, and the extent and temperature condition of constructions (buildings, tunnels, sewers, etc.). The relevant hydrothermal properties of soils are their thermal conductivity and thermal capacity. The former can be determined by a thermal response test. While water exhibits an extremely high thermal capacity, soils are neither good insulators nor highly conductive materials. Properties depend to a high degree on the water content. Hydrogeological and hydrothermal investigations required to furnish the relevant parameters are further discussed in chapter 6.

1.3 TECHNICAL SYSTEMS

1.3.1 Heat pumps

A **heat pump** is a technical device that can transfer thermal energy from a "cold" reservoir (a fluid, e.g., air or water mass) at a lower temperature level to a "hot" reservoir with a higher one (again, e.g., air or water) by using an amount of external energy (electrical or fuel energy, waste heat). An overview on ground-source heat pumps is given in Lund et al. (2004). The two main types are the vapor compression and the absorption cycle heat pumps.

A **vapor compression heat pump** (Figure 1.7) consists of a compressor, an expansion valve, and two heat exchangers referred to as an evaporator (heat source) and a condenser (heat sink). The most common type of heat pump is the vapor compression heat pump powered by electricity. Energy is used to mechanically compress the working fluid (vapor). There exist also vapor compression heat pumps, which can be driven by a combustion engine using natural gas or biogas or by directly using solar energy or geothermal energy (Brenn et al. 2010). **Absorption cycle heat pumps**, on the other hand, are thermally driven. Absorption systems utilize the property of liquids or

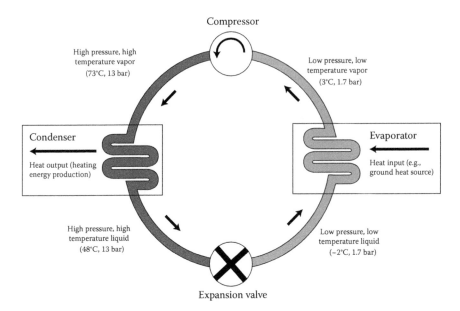

Figure 1.7 Compressor heat pump cycle (schematic).

salts to absorb vapor of the working fluid thus generating heat. Compression of the working fluid is achieved thermally in a solution circuit, which consists of an absorber, a solution pump, a generator, and an expansion valve. Some additional electricity is needed to run a pump. There exist also various hybrid systems. Common to both types is that they use closed cycles with a working fluid. Heat transfer from heat sources and to heat sinks is usually accomplished with the help of heat exchangers. Information can be obtained, for example, at the IEA OECD Heat Pump Centre.

The total heat flux of a heat pump is, in principle, the sum of the heat flux extracted from the heat source and the heat flux needed to drive the cycle. The performance of a heat pump is represented by the **coefficient of performance** (COP). It is defined as the ratio of useful energy delivered and the total external energy consumed by the heat pump. For an electrical compression heat pump, the COP equals the heat output of the condenser versus the electrical input to the compressor. It can be confronted with the **ideal performance of a Carnot cycle:**

$$\varepsilon_{\text{Carnot}} = \frac{T_{\text{hot}}}{T_{\text{hot}} - T_{\text{cold}}}$$

(1.2)

which is defined as the ratio of heat delivered by the heat pump and the energy used by the compressor. It can be considered as a maximum performance, which cannot be exceeded. Typically, a fraction of about 50% of the theoretical ε_{Carnot} can be achieved by a heat pump in practice. The performance is dependent on the heat pump itself (efficiency of heat exchangers, losses in compressor, etc.) and on the temperature difference between the low-temperature medium and the high-temperature medium side. Nevertheless, it becomes clear from Equation 1.2 that a high temperature of the low-temperature medium is more effective. Therefore, ground-source heat pumps and GWHPs usually have a higher COP than air-source heat pumps. This result was demonstrated by, for example, Urchueguía et al. (2008) in their field study in Spain.

Another definition is the **primary energy ratio** (PER), which relates the heat delivered by the heat pump to the primary energy used to generate the external energy supplied. Basically, it is COP times the corresponding efficiency factor. Since COP is often determined for specific operating conditions, the long-term average value in practice is usually smaller. Therefore, a relevant number is the **seasonal performance factor** (SPF), which is an average value over the year. Typical SPF values of GSHP systems are currently around 3 to 4 (Bayer et al. 2012). Miara et al. (2011) evaluated the heat pump efficiency in real-life conditions and found an average SPF value of 3.9 for the 56 GSHP systems studied in Germany. Lund et al. (2011) reported an average COP of 3.5 for GSHP systems. This means that the heat delivered by the heat pump is 3 to 4 times the external energy supplied.

1.3.2 Closed- and open-loop systems

In the thermal use of shallow underground systems, the underground (rock, dry, or unsaturated or saturated soil) or pumped groundwater acts as heat source or heat sink for a heat pump. Various systems exist (Lund et al. 2004; Florides and Kalogirou 2007) for the thermal use of the underground in connection with heat pumps. On one hand, there are ground-coupled **closed-loop systems** (Figure 1.1a) in connection with a heat pump. On the other hand, there are **open systems** with groundwater extraction and injection wells (Figure 1.1b), with or without heat pump. The type chosen in a specific application depends on the operational mode of the system, the soil and rock type, the land available, the groundwater situation, as well as economic and regulatory issues.

In a **closed-loop system**, a closed loop of a pipe (typically high-density polyethylene plastic, or also copper pipe) is installed underground horizontally (at a depth of 1 to 2 m) or vertically (down to a depth of a few tens to about 400 m), acting as a heat exchanger. An environmentally safe heat carrier fluid, for example, a water-antifreeze solution, is circulated through the pipes to collect heat from the ground in winter and/or to inject heat to the ground in summer. The typical **horizontal alignment** of the tube consists of a serial or a parallel connection of pipe sections. According to Florides and Kalogirou (2007), a typical horizontal loop is 35–60 m long for 1 kW of heating or cooling capacity. In horizontal systems, the thermal recharge is provided essentially by solar energy entering the soil surface, which is essentially induced heat from the soil surface. A special variant of horizontal collectors is the direct exchange system. Instead of using plastic pipes, soft copper tubing directly transfers energy between ground and refrigerant. Circulating the heat pump refrigerant instead of the heat carrier fluid in the ground saves one energy transfer step, and the refrigerant temperature is closer to the ground temperature. This is promoted by the copper tubing, which has much higher heat conductivity than plastic. This direct technology raises the SPF by about one unit in comparison to standard practice, however, at the expense of higher installation costs and environmentally critical copper use. Further details on horizontal closed-loop systems can be found in Banks (2008). In **vertical BHEs,** typically plastic pipes (polyethylene or polypropylene pipes) are installed in the borehole. The remaining space is typically filled with a grouting material (backfilling, e.g., concrete–bentonite mixture), which ensures good contact between pipe and undisturbed ground and reduces the thermal resistance. In hard rocks, however, grouting is not needed for stabilization, and if no hydrogeological concerns are raised, no backfilling is applied. Such practice is common in Scandinavian countries such as Sweden. Frequently used types of BHEs are the U-pipe configuration or the coaxial pipe configuration with two concentric tubes (Figure 1.8). In a U-pipe configuration, a pair of straight pipes is connected with a U-bend pipe. One or two U-pipe configurations are typically installed (Figure 1.8a and b). Further configurations exist like the spiral coil or helical configuration. BHEs can be conceived as single systems or as arrays of several to many units (de Paly et al. 2012).

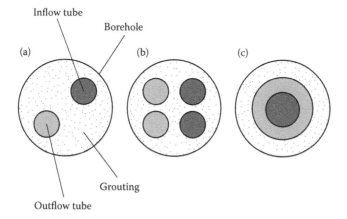

Figure 1.8 Typical BHEs: (a) single U-pipe BHE; (b) double U-pipe BHE; and (c) coaxial BHE.

An **example** is the heat storage project at the Science City Campus (Hönggerberg) of ETH Zurich (Switzerland, Figure 1.9). According to this project, waste heat from the buildings will be dynamically stored underground in summer, with the help of a total of 800 vertical BHE units, and reused in winter for space heating as well as process heat. Two plots of 100 and 130 heat probes have been installed so far and have been operating successfully since 2012 (ETH Life 2012). The vertical probes consist of plastic pipes. They are 200 m long and are arranged in regular intervals of 5 m each horizontally. The local underground consists of unsaturated or quasisaturated moraine material and molasse rock (sandstone, marl, and conglomerates). The BHEs are connected to a pipe network of the buildings. Space heating is performed in winter with the help of heat pumps using the underground heat of the heat storage. At the end, a total of about 4 millions m^3 of underground volume will be utilized, leading to an expected thermal yield of about 15 GWh per year. **Another example** of borehole thermal energy storage is the facility at the University of Ontario, Institute of Technology (Dinçer and Rosen 2011).

A special combination of closed systems consists of **heat exchangers in geotechnical constructions**. These are foundation constructions like piles, plates, etc., which are equipped with built-in heat exchangers (SIA 2005).

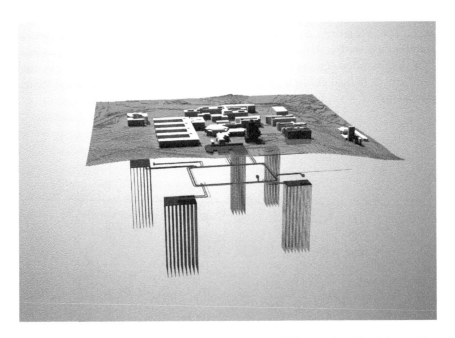

Figure 1.9 **(See color insert.)** Heat storage project (schematic) at the Science City Campus (Hönggerberg) of ETH Zurich (Switzerland). (Courtesy of ETH Zurich, Abteilung Bau, 2011.)

They are typically applied in building constructions under difficult soil conditions, like silt, fine sand, marl, mud, unconsolidated deposits, etc. Often these are fine-grained soils. The tight relationship to the building makes them suitable for combined space heating and cooling with seasonal heat charging and extraction. It is indispensable for these systems that the thermal recharge is guaranteed by proper design in order to avoid excessive cooling or heating of the ground. Particularly, soil freezing has to be prevented in order to preserve the geotechnical functioning of the construction.

The **open-loop system** uses pumped groundwater directly in the heat exchanger and then discharges it to the aquifer, to a stream or lake, or on the soil surface (e.g., infiltration or irrigation), depending upon local regulations. A typical configuration consists of a **well-doublet scheme** in a shallow aquifer with extraction well, which pumps water to the heat exchanger of the heat pump (Figure 1.10). The cooled or warmed water is injected back, either in an injection well or via an infiltration facility (pond, pit, ditch, gallery, well), where water infiltrates toward the groundwater table. Care has to be taken that the distance between extraction well and infiltration facility or injection well is large enough to prevent a hydraulic short circuit, thus avoiding the recycling of cooled or warmed water, respectively. Configurations with several pumping wells and injection facilities are also in use. It is further possible to pump groundwater and to discharge it to a surface water body. The depth of the wells is typically less than 50 m.

Further schemes exist. For example, in the **standing column well**, water is pumped from the bottom of the well to the heat pump. The injected water is percolated through a gravel pack in the annulus of the well in order to absorb heat. Standing column wells are typically 15 cm in diameter and may be as deep as 500 m (Florides and Kalogirou 2007). A detailed description of standing column wells can be found in Banks (2008). A related concept consists of the **vertical double well** after Jacob (Banks 2008). It

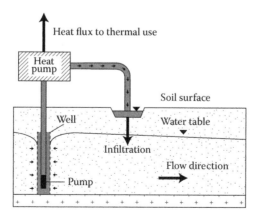

Figure 1.10 Well doublet with pumping well, heat pump, and infiltration facility.

consists of inflow and outflow sections of a well, which are separated by packers and a pump.

1.4 ENERGY DEMAND AND ENERGY PRODUCTION

Basic information needed for the design of technical installations for thermal use of buildings, settlements, and industrial installations is the seasonal energy demand. This comprises both heating and cooling energy. In the discussion about energy demand in the context of heat pumps, it is important to define the **temperature range**, since it affects the selection of the heat pump system:

- Low energy space heating uses water temperatures in the range of about 35°C to 55°C.
- Hot water for domestic use has temperatures in the range of 60°C to 70°C.
- Process water for industry is supplied at the requested temperature (heating or cooling).
- Space cooling requires coolant temperatures in the range of 22°C to 25°C.

A schematic seasonal energy demand profile is shown in Figure 1.11.

Bayer et al. (2012) presented an overview on the current **energy demand** for space heating for selected countries. Except in North European countries, the relative portion of ground-source heat pumps and GWHPs is still small and often supplies less than 1% of the total energy demand.

An overview of the **energy production by ground-source heat pumps in shallow underground systems** (Bayer et al. 2012) is given in Table 1.1 for selected countries (status in 2008). For comparison, the total heating

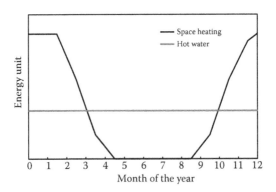

Figure 1.11 Seasonal energy demand (schematic).

Table 1.1 Energy production by GSHPs in shallow underground systems and heating demand in 2008 in European countries

	GSHPs (TJ)	Heating demand (TJ)	Population ×10⁶
Austria	3,229	209,000	8.2
Belgium	495	295,000	10.4
Czech Republic	927	186,000	10.2
Denmark	1,859	173,000	5.5
Finland	9,852	139,000	5.2
France	7,784	1,268,000	63.7
Germany	11,237	2,044,000	82.4
Italy	1,157	816,000	58.1
Netherlands	1,080	298,000	16.6
Norway	8,588	91,000	4.6
Poland	1,193	615,000	38.5
Slovenia	387	30,000	2.0
Spain	408	286,000	40.4
Sweden	52,251	257,000	9.0
Switzerland	7,403	138,000	7.6
United Kingdom	783	1,152,000	60.8

Source: After Bayer, P. et al., *Renewable and Sustainable Energy Reviews* 16(2): 1256–1267.

demand and the population are presented as well. For Switzerland, the total energy demand in 2010 was 911,550 TJ, the energy production by closed ground-source systems was 5321 TJ, and the energy production by open systems was 737 TJ (FOE 2011).

In practice, **back-of-the-envelope** calculations are often useful for first tier screening. A common approach is to predict the specific heat extraction rate $q_{tb} = J/H$ of a closed-system borehole depending on operating hours and ground conditions (Table 1.2). By also assuming that the SPF can be anticipated, the required total borehole length H is estimated for an annual energy demand E and operating time t:

$$H = \frac{E - \dfrac{E}{\mathrm{SPF}}}{tq_{tb,given}} \tag{1.3}$$

As an example, consider the heating of a single family house by a closed system with SPF = 4.25. The annual demand of the house is 21,350 kWh (~2000 kg of fuel oil). The system is operated for $t = 2160$ h during the heating season with a specific heat extraction of $q_{tb} = 50$ W m^{-1}. The question is, what is the needed installation depth H of the BHE to supply the required heat demand? The primary power consumption of the heat pump is governed

Table 1.2 Specific heat extraction rates for single closed system BHEs (W m⁻¹) of 40–100 m length (only heating) for given total annual operation times

Rock and soil material		*VDI 4640* (VDI 2001a)	*VDI 4640* (VDI 2001a)	*MCS* (2011)	*Numerical Simulation* (Erol 2011)
		1800 h/a	*2400 h/a*	*2400 h/a*	*2400 h/a*
Unconsolidated soils (dry conditions)	Sand	<25	<20	n. a.	38–48
	Gravel	<25	<20	n. a.	34–43
Unconsolidated soils (saturated)	Clay	35–50	30–40	22–35	27–32
	Loam (silt)	35–50	30–40	22–35	32–39
	Sand	65–80	55–65	23–46	38–49
Unconsolidated soils (with high groundwater flow)	Sand	80–100	80–100	n. a.	55–82
	Gravel	80–100	80–100	n. a.	67–114
Sedimentary rocks	Sandstone	65–80	55–65	22–55	40–52
	Limestone	55–70	45–60	n. a.	32–39
Magmatic rocks	Basalt	40–65	35–55	n. a.	32–39
	Granite	65–85	55–70	32–48	33–40
Metamorphic rocks	Gneiss	70–85	60–70	n. a.	37–48

by the SPF, and related to the heat demand we obtain 21,350 kWh/4.25 = 5024 kWh. The fraction of geothermal energy thus is 21,350 kWh – 5024 kWh = 16,360 kWh. With the given operation time, the required power of the heat pump results in 16,360 kWh/2160 h = 7559 W. This means, with the given specific heat extraction, a borehole length of 7559 W/50 W/m = 151 m is needed. However, this is only a very approximate value, and the assumed value of the specific heat extraction is uncertain and governed by the hydrogeological, geothermal, and technological conditions. Apart from this, the value of the SPF is not only determined by the efficiency of the heat pump but also strongly influenced by the operating conditions.

Specific heat extraction rates are given by the VDI 4640 guidelines (VDI 2001a) and are commonly used in practice. For these values, however, no quantitative basis is provided. The values are pertinent for double-U-tube or coaxial pipes and for boreholes of 40–100 m length. In comparison, a detailed numerical simulation with a range of different scenarios was conducted by Erol (2011) for typical rock and soil in Germany, given realistic ranges of thermal conductivity and heat capacity. Resulting values of the specific heat extraction rates after 30 years of operation (at 2400 h year⁻¹) are comparable to VDI values for some unconsolidated sediments such as clay and loam, but are higher for sand. For most other scenarios, the computed ranges are below the VDI values. One reason could be the role of groundwater flow, which means additional advective heat provision and thus higher specific heat extraction rates. For the VDI values, this may be accounted for, whereas the simulation was done without. In comparison, as a specific case of unconsolidated material, soils with "high groundwater flow" are distinguished by the VDI with similar values as the simulated ones (seepage velocity of 0.5–5 m day^{-1}). Groundwater flow does not only increase the specific heat extraction rate; it also promotes regeneration. As a result, typical performance decline, reflected in the simulation by decreasing heat extraction rates, is mitigated (Erol 2011). Table 1.2 also lists values taken from a British guideline (MCS 2011), which are comparable to those ranges obtained by simulation.

1.5 MANAGEMENT OF UNDERGROUND RESOURCES

The thermal use of soil and groundwater resources has to be embedded in a holistic view of the management of the water resource. Soil water and groundwater environments are, in principle, **part of an ecosystem with related ecosystem services**. The **management of groundwater resources** essentially comprises its use for **water supply for domestic and industrial use**, as well as the **thermal use**. While questions of the impact of the thermal use on groundwater quality and ecology are treated in Section 1.6, specific questions related to the management of underground resources are briefly discussed here. One management aspect is the **seasonal operation of technical installations for the thermal use of the underground**. Another management aspect is the **possible antagonism between water supply and thermal use**.

1.5.1 Seasonal operation of technical installations

The **seasonal operation of technical installations** for the thermal use of soil and groundwater can be conceived as

- Solely extracting heat from the underground or groundwater, that is, for space heating in winters
- Solely injecting heat into the underground or groundwater, that is, for space cooling in summers
- Combined seasonal heat extraction and injection

Technical systems **solely extracting or injecting heat** only work in a satisfactory, long-term manner if thermal recharge or recovery of the cooled or warmed underground space is assured. Otherwise, undesired soil cooling or soil warming, or even soil freezing, occurs. Efficient heat recharge happens if sufficient advective heat transport by soil water or groundwater flow is present. For a typical field situation, the related minimum Darcy velocity is of the order of 1 m day^{-1}. Otherwise, the thermal recovery is restricted to mainly heat conduction from the soil surface. Proper design of the layout is decisive.

Combined seasonal heat extraction and injection with heat storage enable thermal recharge by the system itself. However, the efficiency might be severely reduced in this case by too high influence of advective heat transport due to water flow. If possible, the technical design should pursue thermal layering instead of mixing of warm and cold water.

1.5.2 Water supply and thermal use

Groundwater resources, which are suited for drinking water supply, often receive highest priority in national regulations for groundwater protection (see Section 1.8). Still the **simultaneous use of groundwater resources for water supply and for thermal use** is possible and desired. However, it deserves some attention with respect to the overall energy balance.

Imagine substantial pumping from an aquifer for water supply at temperature T_0 with the water being delivered to domestic users and industrial plants and installations in buildings. Depending on the residence time of the water and the type of technical installation in the buildings (e.g., non-insulated containers and vessels like water toilets, boilers, and reservoirs) and on the prevailing room temperature in the building, a certain energy flux will result (warming or cooling). This energy flux may have to be compensated by the local climate control systems. Now, if simultaneous thermal use of the aquifer is introduced, this results in a change ΔT of the groundwater temperature (warming or cooling) and therefore in the temperature of the delivered water. In this case, the expected temperature of the pumped and delivered water is essentially changed to $T_0 + \Delta T$, and a compensating energy flux in the buildings and installations may be needed. For an aquifer, the temperature of which is decreased, a compensating positive energy flux within the buildings will be needed in winter. This represents an energy loss. Similar considerations are applicable for increase in groundwater temperature for cooling purposes in summer. Therefore, depending on the water and energy fluxes and the temperature difference ΔT, **short-circuit effects and thus a loss of the overall thermal efficiency** may occur. Such effects should be avoided or at least be reduced.

1.6 IMPACT ON GROUNDWATER QUALITY AND ECOLOGY

Water quality can be affected by heat extraction or injection by **directly influencing water temperature**. It can be indirectly influenced by **temperature-dependent physical, chemical, and biological processes** and their interaction.

Most physical properties are temperature dependent (e.g., water viscosity or water density). Temperature can **affect the thermodynamic equilibrium of species and the chemical milieu**. All chemical reactions are, in principle, temperature dependent. Recall the van-'t-Hoff rule, which states that for a temperature rise of 10 K, the velocity of chemical reactions doubles. Lowering the temperature can change the equilibrium constants of dissolved minerals and gases. A consequence can be, for example, increased solubility of carbon dioxide and a subsequent increase in carbonate hardness. An increase in temperature can lead to increased solubility of minerals or to increased growth of undesired bacteria. Furthermore, undesired desiccation of soils may occur, thus leading to undesired soil cracks. Temperature change may **affect the self-purification ability of soils and groundwater systems**. **Plant growth in ecosystems including agriculture** is sensitive to temperature changes. The **effect of thermal energy discharge on shallow groundwater ecosystems** was investigated by Brielmann et al. (2009). For their investigation site in Germany, they found no likely threat to ecosystem functioning and drinking water quality by the thermal use of the aquifer. **Shallow soils** may be strongly thermally affected by horizontal closed heat exchanger systems. Therefore, care has to be taken in the design of such systems in order to avoid negative effects. The same holds true for the neighborhood of vertical closed systems.

Water quality can further be negatively affected by **pollutants** entering soil and groundwater via installations for thermal use, for example, via boreholes or via infiltration facilities. Potential pollutants are the working fluid of heat pumps, the heat carrier fluid in closed systems, or any further pollutant entering soil and groundwater via BHE, pumping well, or infiltration facilities. This concerns mainly GWHP systems and those ground-source heat pump systems, which are situated in groundwater supply areas.

1.7 GEOTECHNICAL ISSUES

Due to the development of geothermal systems **malfunctioning, direct damages** of the systems or even **third party damages** may occur. Causes for such damages are numerous and might be related to the technical heat pump systems, site-specific conditions of the subsurface, and/or the development by the geothermal systems, such as drilling of the wells or installation of BHEs. Bassetti et al. (2006), for example, investigated causes of damages by ground-source heating pump (GSHP) systems in Switzerland. They concluded that many causes are related to the incorrect design of GSHP systems, resulting mainly in malfunctioning of the GSHP system. Blum et al. (2011), who studied 1100 GSHP systems in Germany, also showed that subsurface characteristics are often not adequately considered during the planning and design, which results in an undersizing or oversizing of GSHP systems.

In addition to such deficiencies during the planning and design, **geotechnical issues** such as **subsidence** or **uplift** and resulting damages may occur due to the installations of geothermal wells or BHEs. For example, **inadequate backfilling** of the annular space between the BHE and the borehole wall may cause severe damages due to the rise of artesian groundwater or the artificial hydraulic connection between aquifers. The latter, for example, resulted in severe third party damages due to land subsidence and uplift in South Germany. In 2007, a heave of more than 10 cm per year triggered by anhydritic swelling caused a severe damage in the historic town of Staufen (Goldscheider and Bechtel 2009; Sass and Burbaum 2010; LGRB 2010, 2012). The total damage of more than 250 houses including the historical town hall (Figure 1.12) is currently estimated to be around 50 million euros. The site is currently remediated by pumping groundwater from the lower Triassic limestone aquifer.

In contrast to the damage in Staufen, one year later in 2008, an artificial hydraulic connection between two aquifers due to the installation of a GSHP system caused a local subsidence in the town of Schorndorf (Germany), and particularly, one grammar school house showed numerous cracks and one spring fell dry. In 2009, the BHEs were successfully remediated by overcoring and grouting, using a concrete–bentonite suspension. The costs incurred in this case were only about 300,000 euros. Both cases, however, clearly demonstrate the potential damages that might be triggered due to geothermal installations. Such third party damages due to installations of GSHP systems in the state of Baden-Württemberg (southwest Germany) are site specific and mainly occurred by drilling in the Triassic Gipskeuper Formation and the Triassic limestones (Muschelkalk). Furthermore, both acute damages in the towns of Staufen and Schorndorf were caused by inadequate backfilling of the BHE, and hence damages could have been avoided by adequate and rapid on-site measures. After these incidents, a **technical guideline** for the quality assurance for GSHP systems was introduced by the Environmental Ministry of Baden-Württemberg (Germany) in 2011 (LQS 2012).

Nevertheless, future damage due to the installations of a geothermal system can never be ruled out; however, it can be comprehensively addressed by improved locally adopted license systems (e.g., Butscher et al. 2011) and **quality standards and quality assurances** (QS/QA). On the other hand, such damages might also be only locally restricted, site- and country-specific. For example, in Scandinavia, boreholes are typically drilled in crystalline rocks and are not grouted, leaving the space between the BHE pipes and the borehole wall filled with groundwater (e.g., Gustafsson et al. 2010). Under such conditions, the geotechnical and third party damages, which occurred in southwest Germany, are not feasible.

Figure 1.12 Damaged town hall of the historic town of Staufen, which was triggered by geothermal installations that resulted in the swelling of clay-sulfate rocks in the Gipskeuper Formation. (Courtesy of Christoph Butscher, 2013.)

1.8 REGULATORY ISSUES

As a rule, groundwater resources, which are suited for **drinking water supply, receive highest priority in groundwater protection**. Therefore, parts of the regulations concerning thermal use are based on groundwater protection principles, which primarily concern water quality issues (e.g., Stauffer 2011). Of course, quality can be affected by temperature, directly and indirectly. **Drinking water quality targets** may comprise temperature ranges, which can be exceeded or undercut by thermal use (warming and cooling). Water quality is indirectly influenced by **temperature-dependent physical, chemical, and biological processes** and their interaction. Therefore, regulations **limit the temperature change** caused by thermal use with respect to the initial, unaffected state. It is sometimes argued that admissible temperature changes should be smaller than climatically caused deviations from seasonal temperature variations. Some uncertainty exists with respect to the initial thermal state, which is the basis for evaluating the permissible temperature change. Regulations may also concern the energy abstraction rate (energy load).

Water quality can further be negatively affected by **pollutants** entering soil and groundwater via installations for thermal use. Therefore, regulations may concern technical and operational requirements in order to minimize or limit pollution.

Further regulations concern **soil and ecosystem protection**. Plant growth and agricultural activities should not be deteriorated by thermal use. These requirements affect mainly horizontal closed-loop systems. Some regulations limit the admissible temperature range.

Regulations also concern the heat pumps. One of the most important items concerns the choice of the **working fluid of heat pumps**. Working fluids principally have to fulfill the requirement of protecting the atmospheric ozone layer. Moreover, they have to fulfill the requirements of soil and groundwater protection.

Thermal use of soils and groundwater represents the **exploitation of a renewable resource**. This may affect neighboring stakeholders. Any thermal use results in temperature anomalies, which propagate within the underground. Since the temperature change should not exceed a predescribed limit, the thermal use of an individual abstractor can be strongly affected by upstream or neighboring users. Therefore, regulations may concern the **prevention of disadvantage** by, for example, limiting the distance of open and closed systems to neighboring property lines, wells, and surface water bodies.

Regulations may also prescribe **thermal monitoring** of groundwater resources and possibly of soils. Long-term thermal monitoring allows the surveillance of efficient use and of measures to ensure the quality standards.

Main regulatory issues and recommendations may comprise the following items, depending on the responsible agency:

- **Need for a concession** or license for thermal use of underground space or groundwater issued by the responsible agency
- Definition of permissible maximum modification of the seasonal ground and groundwater temperature
- Definition of **minimum** (e.g., 5°C) and **maximum** (e.g., 25°C) **groundwater temperature**
- Prevention **of disadvantage** to neighboring property owners by thermal use
- Prevention of groundwater pollution by thermal use systems
- Licensing of **working fluids** for heat pumps
- **Thermal monitoring** of groundwater resources

Haehnlein et al. (2010) presented an overview on the **international legal status** of the thermal use of shallow aquifer systems. They showed that the legal situation is quite diverse. Extensive national regulations exist only in a few countries such as Denmark or Sweden. European countries are, in general, more regulated. In addition, guidelines and technical recommendations exist in several countries, somehow defining the state of the art. A wide range for minimum required distances of thermal installations from property lines was observed (5–30 m). The same holds true for the temperature limits for groundwater. Specific ecological and environmental criteria are rarely reported. The authors observed that the highest inconsistency occurred in values for the acceptable maximum temperature change for groundwater, which is 3 K in Switzerland, 6 K in Austria, and 11 K in France.

On the **European Community level**, the use of energy from renewable sources is promoted by the directive 2009/28/EC (EC 2009). The directive includes the thermal use of shallow geothermal systems. An overview of the current European regulations for geothermal energy can be found in the EGEC (2007) report. Accordingly, the relevant national legislation is spread throughout the mining, energy, environmental, water management, and geological acts, sometimes in a contradicting way, and the licensing authority framework for geothermal facilities is rather complex in most countries.

In the following, a few **examples of national regulations** on the use of shallow geothermal energy are described.

1.8.1 Swiss regulation

According to the **regulation of Switzerland** (FOEN 2004), the power of allocating water resources, including groundwater, is basically a privilege of the cantons (provincial level). Based on federal and cantonal law, rights can be further transferred to beneficiaries (communities, corporations). In general, these authorities decide upon the permission or license for thermal use of groundwater resources by applicants (communities, companies, private persons). Based on federal and cantonal law, soils can generally be thermally used, for example, by horizontal closed-loop systems by property owners. On the other hand, BHEs and installations with pumping well and infiltration device for thermal use are, in general, not permitted in aquifers that are suitable for drinking water supply. Therefore, priority is clearly given to drinking water supply. Exceptions can be made by the authority, which issues the corresponding concession or license. Measures have to be taken in order to prevent groundwater pollution, for example, by leaking heat exchangers. Moreover, it has to be ensured that no further pollutants can enter the subsoil.

Based on federal law, the thermal use of an aquifer may not change the seasonal groundwater temperature by more than 3 K with respect to a situation with "natural" or "close to natural" conditions. However, in the immediate neighborhood of infiltration facilities, the temperature change may be higher, but has to reach the 3 K limit within a distance of maximally 100 m from the infiltration facility. The limit has to be fulfilled under consideration of all installations for thermal use within the aquifer. Outside of usable groundwater resources, the thermal use with closed heat exchanger systems (horizontal or vertical loops) is generally acceptable. Further restrictions can be formulated by the cantons. The regulation of admissible cooling and warming of groundwater directly limits its thermal use.

The definition of **natural conditions** represents a challenge since such conditions do not exist anymore. They can be approximated at best. We have to keep in mind that mainly in urban areas, with relatively small depths to groundwater, the groundwater temperature is often increased due to the settlements, factories, and technical installations, as well as increased air temperature, and therefore also mirrors the typical heat island effect. Thermal effects are present, in principle, also for agricultural and horticultural areas. These anthropogenic temperature changes have to be taken into consideration in the thermal management of aquifers (Figure 1.13). The limitation of temperature change with respect to "natural" conditions is mainly meant to limit undesired physical, chemical, and biological effects on the groundwater quality.

The Swiss Federal Government issued the implementation tool "Use of heat from soil and subsoil" (in German and French; FOEN 2009). It is intended to ensure harmonization of the approval practice of the cantons for shallow geothermal heat probes, GWHPs, soil recorders, and geothermal energy cages and piles in Switzerland. It also defines the necessary conservation measures, on the basis of the environmental protection legislation. Technical standards exist for the installation of shallow geothermal heat probes (SIA 2010).

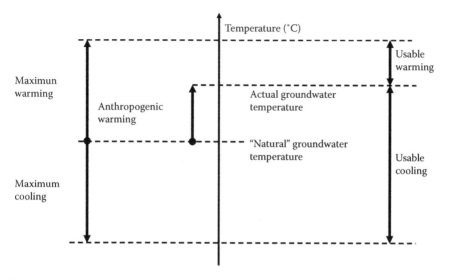

Figure 1.13 Usable warming and cooling range in regulations, which allow maximum deviation from "natural" conditions (schematic).

Concerning the **installation of vertical geothermal heat probes,** the Swiss Federal regulation (FOEN 2009) lists the regions where such installations are prohibited. These regions concern (1) aquifers, which are suitable for drinking water supply, (2) geological formations with high hydraulic conductivity and therefore large flow velocities, like karst formations or highly conductive fractured rock formations, and (3) potential landslide areas. Furthermore, situations are listed where the installation can be approved after detailed investigations, like installations in confined aquifers or multiaquifer systems, or the existence of contamination at the site, or the existence of highly mineralized groundwater. Exceptions are possible and need specific investigations and the approval by the cantonal authorities. The main motivation is to prevent groundwater pollution.

The required **steps for the design of vertical geothermal heat probes** for heating purpose according to SIA (2010) are as follows:

1. Determination of the mean soil surface temperature of the site
2. Determination of the thermal parameters (thermal conductivity)
3. Determination of the standard specific performance of the geothermal heat probe, depending on the thermal conductivity (diagram)
4. Evaluation of the standard annual hours of operation, depending on elevation (Switzerland)
5. Evaluation of annual energy demand of the system
6. Evaluation of the total annual hours of operation of the system

7. Evaluation of standard length of probe (diagram)
8. Correction of length of probe for given probe configuration and total annual hours of operation
9. Correction of length of probe for mean soil surface temperature
10. Evaluation of pressure loss in the pipe of the probe
11. Determination of the total length of the probe

The specifications are valid for Switzerland. With these requirements, the minimum temperature of 0°C/–3°C (outlet/inlet) of the refrigerant fluid of the probe after 50 operating years can be achieved. Another important requirement is to guarantee a professional and tight backfilling of the space between geothermal heat probe and underground material with practically impermeable grout material. This should prevent a hydraulic connection of aquifer layers with differing water quality and also prevent the penetration of pollutants into the subsurface from the soil surface or from leaky tubes.

The Swiss Society of Engineers and Architects (SIA) issued a series of documentations, which provide basic information on the design of installations for seasonal energy storage (SIA 1989, 2003), the use of shallow underground systems for space heating (SIA 1996), and the use of foundation constructions equipped with built-in heat exchangers (SIA 2005). They represent the technical state of the art.

We may look at the **regulations of the Canton of Zurich as an example**. The use of energy from underground systems and groundwater is described in a planning support brochure (AWEL 2010). Besides licensing, the local regulations mainly concern BHEs. The location of BHEs within a property is generally unconstrained by regulations with exceptions. These are as follows:

- If the BHE is located less than 3 m from the property line, the agreement of the neighbor is needed.
- If the BHE is located within a communal restricted area (mainly infrastructure), the agreement of the community is needed.
- If the BHE is located close to a railway line or tunnel, the agreement of the railway company is needed.
- BHEs close to surface water bodies are not allowed within a distance of 20 m from the shoreline. For small creeks, the distance is smaller.

The grouting material of BHEs has to fulfill specific requirements concerning the material composition (a bentonite cement water mixture). Conformity of the materials with water protection requirements has to be observed. Furthermore, a long-term (50 years) assessment of the temperature development has to be provided (if heat extraction capacity is larger than 100 kW).

1.8.2 Austrian regulation

According to the **Austrian standards** (ÖWAV 2009), drinking water supply is clearly prioritized. Thermal use within well-head protection zones is not allowed in protection zone I (the immediate zone around the extraction well) and is limited in protection zones II and III (zone II according to bacteriological protection with groundwater residence time of 60 days and zone III for chemical pollution prevention). Moreover, thermal use of confined aquifers is not allowed. Existing entitlements of thermal use have to be respected.

Based on the observation that temperature in shallow aquifers at a depth of 7 m below ground surface typically ranges between 7°C and 12°C, the temperature at the location of the water injection from thermal use should not fall below 5°C and not exceed 20°C. The maximum warming or cooling is limited to 6 K at the location of the water injection, compared with the existing temperature. Water should be extracted from deep sections of the aquifer, if possible, whereas reinjection should preferably be done at shallow depths.

In BHEs, as well as shallow horizontal heat exchangers, the temperature of the heat carrier fluid should not exceed 30°C (for cooling purposes). The mean temperature (average of inflow and outflow temperature) of the heat carrier fluid should not fall below –1.5°C after reaching equilibrium conditions (expected after 5 to 50 years). Temperature in shallow horizontal heat exchangers should not lead to freezing. Further requirements concern the heat carrier fluid and the working fluid of heat pumps, as well as the design of systems for thermal use.

1.8.3 British regulation

Recommendations and regulations concerning the thermal use of groundwater are based on policy and practice of groundwater protection (EA 2008). **Key issues** are as follows:

- The risk of the pipes or boreholes creating undesirable connections between rock or soil layers. This may cause pollution and/or changes in groundwater flow and/or quality.
- Undesirable or unsustainable temperature changes in the aquifer or dependent surface waters.
- Pollution of water from leaks of polluting chemicals contained in closed-loop systems.
- Pollution of water from heat pump discharge from an open-loop system that contains additive chemicals.
- Impacts of reinjection of water from an open-loop system into the same aquifer, both hydraulic and thermal, as well as any water quality changes induced.

- The potential impact of groundwater abstraction for ground-source heat systems on other users of groundwater or surface water.

Requirements can be summarized as follows:

- It is strongly recommended that GSHP systems are operated sustainably.
- Developers are expected to undertake appropriate prior investigations of the planned systems.
- Drilling through contaminated ground poses a significant risk of groundwater pollution.
- Where it is necessary to prevent pollution of controlled waters, the agency serves a notice under the Groundwater Regulations or Water Resources Act to control the activity.
- A permit is not required for closed-loop systems. It is strongly recommended to not use hazardous substances.
- An abstraction license is required unless the abstraction rate is below the threshold for license control, which is currently 20 m^3 day^{-1}.
- Discharge consent is required unless it can be shown that no deterioration in groundwater quality is caused and that no significant change in ground or groundwater temperature occurs.
- If the GSHP system adversely affects existing systems or other legitimate use of groundwater, the operation of the systems has to be modified.

1.8.4 German regulation

In Germany, the mining law (Bundesberggesetz BBergG) states that shallow geothermal energy belongs to the 16 federal states, and, generally, these states provide individual guidelines. These serve as the basis for adapting local regulations and permission procedure, and they are mainly focused on closed systems. We find suggested minimum distances between 5 and 10 m between two BHEs and 3 and 5 m between property line and BHE (Haehnlein et al. 2010). In addition to the regulatory framework by the states, as fundamental general and technical guidelines, the VDI 4640 (VDI 2001a,b, 2004, 2010), VBI-guideline (VBI 2008), and for quality assurance, state-specific standards (e.g., LQS 2012) are consulted. These include suggestions on how to restrict the temperature difference and maximum temperature for heating and cooling of the heat carrier fluid and for the groundwater. If groundwater is concerned, then the federal Water Act is applicable (Haehnlein et al. 2011). It states that detrimental changes in physical, chemical, and biological characteristics must be avoided. A critical point, however, is that detrimental changes are not further specified. A unique guideline for groundwater injection and extraction systems for thermal use is available from the state of Baden-Württemberg (Bauer et al. 2009). This includes an analytical equation describing heat conduction, which can be applied to approximate thermal conditions evolving in an aquifer with operating wells.

1.9 CHALLENGES RELATED TO DESIGN AND MANAGEMENT

From the discussion above, we can list a few **problems** that are related to design and management of the thermal use of underground space:

- The establishment of closed and open systems for thermal use creates local and regional temperature anomalies of the underground space (rock zone or unsaturated or saturated zones of the aquifer). Regulations may directly refer to expected changes in temperature due to the chosen scheme of thermal management. Therefore, **methods and models are needed to simulate the local and regional temperature development in the underground or aquifer**. Aspects of the investigation can be the short- and long-term behavior, as well as the interference between various installations. The latter is of importance in order to prevent a thermal overuse of the underground space or aquifer and to assess the long-term heat abstraction potential.
- Application of extraction and injection wells in open systems alters both the **thermal and hydraulic conditions** in aquifers. As a consequence, this type of use may compete with installations for freshwater abstraction, and especially in densely populated areas, balanced and concerted management strategies have to be developed.
- In the design of **BHE** systems, an important requirement is to **prevent freezing of the soil and of the circulating fluid**. Therefore, methods and models are needed to simulate the temperature development within and outside of the BHE. Especially the long-term behavior is of importance.
- Additionally to thermal and hydraulic effects, chemical conditions in aquifers change with associated **potential ecological consequences**. Groundwater ecosystems will adapt to thermal, chemical, and hydraulic modifications, which can be of short duration, seasonal, permanent, and even irreversible, slowly evolving or unnaturally abrupt.
- Good characterization of thermally used aquifers is essential to develop efficient management strategies and to make reliable long-term predictions. For this, the repertoire of available hydraulic **field investigation techniques** is extensive. These have to be combined with thermal characterization methods. Still, for being able to predict decades of operation, we also need validation via long-term monitoring programs. We meanwhile have a long-term experience with many applications; however, case studies that provide detailed recorded field and technological performance data are scarce.

1.10 SCOPE OF THE BOOK

The **main objective of the book is to provide and discuss mathematical modeling tools**, which are useful for the design and management of systems making thermal use of underground. Based on the motivation presented in the introduction, the **theoretical foundations** of heat transport in underground systems are recalled, and the essential thermal properties and parameters are reviewed. An overview over a series of **analytical and numerical methods** and models is presented and discussed. The main focus will be the local and regional modeling of flow and heat transport in open as well as closed systems. Using these concepts and models, the long-term operability of thermal systems is discussed. Since any modeling effort has to be combined with an assessment of the hydrogeological–thermal conditions, a series of **field methods** is presented. Finally, **case studies** for locations in Austria, Germany, and Switzerland will illustrate urban thermal energy use as well as heat storage and cooling.

REFERENCES

Allen, A., Milenic, D., Sikora, P. (2003). Shallow gravel aquifers and the urban 'heat island' effect: A source of low enthalpy geothermal energy. *Geothermics* 32, 569–578.

AWEL (2010). *Energienutzung aus Untergrund und Grundwasser (Use of energy from underground systems and groundwater)*. Amt für Abfall, Wasser, Energie, Luft, Canton of Zurich, Zurich, Switzerland.

Balke, J.-D. (1974). Der thermische Einfluss besiedelter Gebiete auf das Grundwasser, dargestellt am Beispiel der Stadt Köln. *gwf Wasser/Abwasser* 115(3), 117–124.

Balke, J.-D. (1977). Das Grundwasser als Energieträger. *Brennst.-Wärme-Kraft* 29(5), 191–194.

Banks, D. (2008). *An Introduction to Thermogeology: Ground Source Heating and Cooling*. Blackwell Publishing, Oxford, UK.

Bassetti, S., Rohner, E., Signorelli, S., Matthey, B. (2006). Dokumentation von Schadensfällen bei Erdwärmesonden. Schlussbericht, EnergieSchweiz.

Bauer, M., Eppinger, A., Franssen, W., Heinz, M., Keim, B., Mahler, D., Milkowski, N., Pasler, U., Rolland, K.M., Schölch-Ighodaro, R., Stein, U., Vöröshazi, M., Wingering, M. (2009). *Leitfaden zur Nutzung von Erdwärme mit Grundwasserwärmepumpen*. Umweltministerium Baden-Württemberg, Stuttgart, Germany, 34 pp.

Bayer, P., Saner, D., Bolay, S., Rybach, L., Blum, P. (2012). Greenhouse gas emission savings of ground source heat pump systems in Europe: A review. *Renewable and Sustainable Energy Reviews* 16(2), 1256–1267.

Blum, P., Campillo, G., Kölbel, T. (2011). Techno-economic and spatial analysis of vertical ground source heat pump systems in Germany. *Energy* 36, 3002–3011.

Blum, P., Campillo, G., Münch, W., Kölbel, T. (2010). CO_2 savings of ground source heat pump systems—A regional analysis. *Renewable Energy* 35, 122–127.

Brenn, J., Soltic, P., Bach, C. (2010). Comparison of natural gas driven heat pumps and electrically driven heat pumps with conventional systems for building heating purposes. *Energy and Buildings* 42, 904–908.

Brielmann, H., Griebler, C., Schmidt, S.I., Michel, R., Lueders, T. (2009). Effects of thermal energy discharge on shallow groundwater ecosystems. *FEMS Microbiology Ecology* 68(3), 273–286.

Butscher, C., Huggenberger, P., Auckenthaler, A., Bänninger, D. (2011). Risikoorientierte Bewilligung von Erdwärmesonden. *Grundwasser* 16(1), 13–24.

Changnon, S.A. (1999). A rare long record of deep soil temperatures defines temporal temperature changes and an urban heat island. *Climate Change* 42, 531–538.

de Paly, M., Hecht-Mendez, J., Beck, M., Blum, P., Zell, A., Bayer, P. (2012). Optimization of energy extraction for closed shallow geothermal systems using linear programming. *Geothermics* 43, 57–65.

Dinçer, I., Rosen, M.A. (2011). *Thermal Energy Storage. Systems and Applications*. Wiley, Chichester, UK.

EA (2008). Groundwater protection: Policy and practice (GP3). Part 4 Legislation and policies. Environmental Agency, UK, Edition 1.

EC (2009). Directive 2009/28/EC of the European Parliament and of the Council 23 April 2009 on the promotion of the use of energy from renewable sources and amending and subsequently repealing Directives 2001/77/EC and 2003/30/EC. European Community.

EGEC (2007). Geothermal heating and cooling action plan for Europe. K4RES-H brochure, *European Geothermal Energy Council*, www.erec.org.

Erol, S. (2011). Estimation of heat extraction rates of GSHP systems under different hydrogelogical conditions. MSc. thesis, University of Tübingen, Tübingen, Germany, 85 pp.

ETH Life (2012). *Erdspeicher in Betriebgenommen (Ground Heat Storage Started Operation)*. Corporate Communications, ETH Zurich, Zurich, Switzerland.

Ferguson, G., Woodbury, A.D. (2004). Subsurface heat flow in an urban environment. *Journal of Geophysical Research* 109, B02402, doi:10.1029/2003JB002715.

Ferguson, G., Woodbury, A.D. (2007). Urban heat island in the subsurface. *Geophysical Research Letters* 34, L23713.

Florides, G., Kalogirou, S. (2007). Ground heat exchangers—A review of systems, models and applications. *Renewable Energy* 32, 2461–2478.

Fluker, B.J. (1958). Soil temperatures. *Soil Science* 86, 35–46.

FOE (2011). Schweizerische Statistik der erneuerbaren Energien. Ausgabe 2010 (Swiss statistic on renewable energy, 2010). Federal Office for Energy, Bern, Switzerland, Swiss Confederation, Bundesamt für Energie.

FOEN (2004). Wegleitung Grundwasserschutz (Guide for Groundwater Protection). Federal Office for the Environment, Bern, Switzerland, Swiss Confederation, Bundesamt für Umwelt.

FOEN (2009). Wärmenutzung aus Boden und Untergrund (Use of heat from soil and subsoil). Federal Office for the Environment, Bern, Switzerland, Swiss Confederation, Bundesamt für Umwelt.

Gallo, K., Hale, R., Tarpley, D., Yu, Y. (2011). Evaluation of the relationship between air and land surface temperature under clear- and cloudy-sky conditions. *American Meteorological Society* 50, 767–775.

Goldscheider, N., Bechtel, T.D. (2009). Editors' message: The housing crisis from underground—Damage to a historic town by geothermal drillings through anhydrite, Staufen, Germany. *Hydrogeology Journal* 17, 491–493.

Gustafsson, A.M., Westerlund, L., Hellström, G. (2010). CFD-modelling of natural convection in a groundwater-filled borehole heat exchanger. *Applied Thermal Engineering* 30(6–7), 683–691.

Haehnlein, S., Bayer, P., Blum, P. (2010). International legal status of the use of shallow geothermal energy. *Renewable and Sustainable Energy Reviews* 14, 2611–2625.

Haehnlein, S., Blum, P., Bayer, P. (2011). Oberflächennahe Geothermie—Aktuelle rechtliche Situation in Deutschland. *Grundwasser* 16(2), 69–75.

Kollet, S.J., Cvijanovic, I., Schüttemeyer, D., Maxwell, R.M., Moene, A.F., Bayer P. (2009). The influence of rain sensible heat and subsurface energy transport on the energy balance at the land surface. *Vadose Zone Journal* 8(4), 846–857.

Landsberg, H. (1956). *The Climate of Towns*. University of Chicago Press, Chicago.

Lee, J.-Y. (2006). Characteristics of ground and groundwater temperatures in a metropolitan city, Korea: Considerations for geothermal heat pumps. *Geosciences Journal* 10(2), 165–175.

LGRB (2010). Geologische Untersuchungen von Baugrundhebungen im Bereich des Erdwärmesondenfeldes beim Rathaus in der historischen Altstadt von Staufen i. Br. Landesamt für Geologie, Rohstoffe und Bergbau Baden-Württemberg (LGRB), Az.: 94-4763//10-563, Freiburg i. Br., 304 pp.

LGRB (2012). Zweiter Sachstandsbericht zu den seit dem 01.03.2010 erfolgten Untersuchungen im Bereich des Erdwärmesondenfeldes beim Rathaus in der historischen Altstadt von Staufen i. Br. Landesamt für Geologie, Rohstoffe und Bergbau Baden-Württemberg (LGRB), Az.: 94-4763//12-2487, Freiburg i. Br., 110 pp.

LQS (2012). *Leitlinien Qualitätssicherung Erdwärmesonden (LQS EWS)*. Ministerium für Umwelt, Klima und Energiewirtschaft (Environmental Ministery), State of Baden-Württemberg, Germany.

Lund, J., Sanner, B., Rybach, L., Curtis, R., Hellström, G. (2004). Geothermal (ground-source) heat pumps—A world overview. *Geo-Heat Centre Quarterly Bulletin* 25(3), 1–10.

Lund, J.W., Freeston, D.H., Boyd, T.L. (2011). Direct utilization of geothermal energy 2010 worldwide review. *Geothermics* 40, 159–180.

MCS (Microgeneration Certification Scheme) (2011). *Microgeneration installation standard: MIS 3005. Issue 3.0*. Department of Energy and Climate Change, London.

Menberg, K., Bayer, P., Zosseder, K., Rumohr, S., Blum, P. (2013). Subsurface urban heat islands in German cities. *Science of the Total Environment* 442, 123–133.

Miara, M., Günther, D., Kramer, T. Oltersdorf, T., Wapler, J. (2011). Heat pump efficiency. Analysis and evaluation of heat pump efficiency in real-life conditions. Report Fraunhofer Institute for Solar Energy Systems ISE, Freiburg, Germany.

Oke, T.R. (1973). City size and the urban heat island. *Atmospheric Environment* 7, 769–779.

ÖWAV (2009). Thermische Nutzung des Grundwassers und des Untergrunds—Heizen und Kühlen (Thermal use of groundwater and underground—Heating and cooling). ÖWAV-Regelblatt 207, Österreichischer Wasser- und Abfallwirtschaftsverband ÖWAV (Guideline 207 of the Austrian Water and Waste Management Association), Vienna, Austria.

Pollack, H.N., Hurter, S.J., Johnson, J.R. (1993). Heat flow from the earth's interior: Analysis of the global data set. *Reviews of Geophysics* 31, 267–280.

Putnam, S.N., Chapman, D.S. (1996). A geothermal climate change observatory: First year results from Emigrant Pass in north-west Utah. *Journal of Geophysical Research: Solid Earth* 101(B10), 21877–21890. doi:10.1029/96JB01903.

Sass, I., Burbaum, U. (2010). Damage to the historic town of Staufen (Germany) caused by geothermal drillings through anhydrite-bearing formations. *Acta Carsologica* 39(2), 233–245.

SIA (1989). Wegleitung zur saisonalen Wärmespeicherung (Guidance for seasonal energy storage). Schweizer Ingenieur und Architektenverband (Swiss Society of Engineers and Architects). Technical documentation D 028.

SIA (1996). Grundlagen zur Nutzung der untiefen Erdwärme für Heizsystems (Basics for the use of shallow geothermal energy for space heating). Schweizer Ingenieur und Architektenverband (Swiss Society of Engineers and Architects). Technical documentation D 0136.

SIA (2003). Energie aus dem Untergrund (Energy form underground systems). Schweizer Ingenieur und Architektenverband (Swiss Society of Engineers and Architects). Technical documentation D 0179.

SIA (2005). Nutzung der Erdwärme mit Fundationspfählen und anderen erdberührten Betonbauteilen. Schweizer Ingenieur und Architektenverband (Swiss Society of Engineers and Architects). Technical documentation D 0190.

SIA (2010). Erdwärmesonden (Ground heat probes). Schweizer Ingenieur und Architektenverband (Swiss Society of Engineers and Architects). Swiss Standards 384/6.

Smerdon, J.E., Pollack, H.N., Cermak, V., Enz, J.W., Kresl, M., Safands, J., Wehmiller, J.F. (2006). Daily, seasonal, and annual relationships between air and subsurface temperatures. *Journal of Geophysical Research* 111, D07101, doi:10.1029/2004JD005578.

Stauffer, F. (2011). Protection of groundwater environments. In: A.N. Findikakis, K. Sato (Eds.), Groundwater Management Practices. IAHR Monograph, CRC Press, Boca Raton, FL, USA.

Taniguchi, M., Uemura, T., Jago-on, K. (2007). Combined effects of urbanization and global warming on subsurface temperature in four Asian cities. *Vadose Zone Journal* 6(3), 591–596.

Taylor, C.A., Stefan H.G. (2009). Shallow groundwater temperature response to climate change and urbanization. *Journal of Hydrology* 375, 601–612.

Turkoglu, N. (2010). Analysis of urban effects on soil temperature in Ankara. *Environmental Monitoring and Assessment* 16969(1–4), 439–450.

Urchueguía, J.F., Zacarés, M., Corberán, J.M., Martos, J., Witte, H. (2008). Comparison between the energy performance of a ground coupled water to water heat pump system and an air to water heat pump system for heating and cooling in typical conditions of the European Mediterranean coast. *Energy Conversion and Management* 49, 2917–2923.

USDA (1999). Soil taxonomy. A basic system of soil classification for making and interpreting soil surveys. United States Department of Agriculture, Natural Resources Conservation Service, Agriculture Handbook No. 436.

VBI (2008). VBI-Leitfaden Oberflächennahe Geothermie, Verband Beratender Ingenieure, p. 59.

VDI (2001a). Verein Deutscher Ingenieure, Blatt 2: Thermische Nutzung des Untergrundes—Erdgekoppelte Wärmepumpenanlagen [Part 2: Ground source heat pump systems], VDI-4640/2.

VDI (2001b). Verein Deutscher Ingenieure, Blatt 3: Thermische Nutzung des Untergrundes—Unterirdische Thermische Energiespeicher [Part 3: Utilization of the subsurface for thermal purposes—Underground thermal energy storage], VDI-4640/3.

VDI (2004). Verein Deutscher Ingenieure, Blatt 4: Thermische Nutzung des Untergrundes—Direkte Nutzung [Part 4: Thermal use of the underground—Direct uses], VDI-4640/4.

VDI (2010). Verein Deutscher Ingenieure, Blatt 1: Thermische Nutzung des Untergrundes—Grundlagen, Genehmigungen, Umweltaspekte [Part 1: Thermal use of the underground—fundamentals, approvals, environmental aspects], VDI-4640/1.

Williams, P.J., Smith, M.W. (1989). *The Frozen Earth. Fundamentals of Geocryology*. Cambridge University Press, Cambridge, UK.

Wu, J., Nofziger, D.L. (1999). Incorporating temperature effects on pesticide degradation into a management model. *Journal of Environmental Quality* 28, 92–100.

Zhu, K., Blum, P., Ferguson, G., Balke, K.-D., Bayer, P. (2010). Geothermal potential of urban heat islands. *Environmental Research Letters* 5, 044002, doi:10.1088/1748-9326/5/4/044002.

Chapter 2

Fundamentals

2.1 THEORY OF WATER FLOW AND HEAT TRANSPORT IN THE SUBSURFACE

The modeling of water flow and heat transport processes in the subsurface has to be based on a **mathematical formulation** of the various processes occurring in a considered spatial domain of the subsurface. Such a mathematical formulation provides a compact description of the relevant processes and includes initial and boundary conditions thus representing a model of the complex reality. Furthermore, it lists all assumptions and simplifications, which are postulated. The considered domain extends horizontally and vertically, the size depending on expected length scales of the processes, as well as on possibilities to formulate proper boundary conditions for the related variables like temperature or infiltration rate. For the modeling of heat transfer in shallow subsurface systems, **the considered domain typically includes the soil surface** where it is feasible to formulate thermal boundary conditions. This means, for the case of shallow unconfined groundwater systems, that, in general, the unsaturated zone (also referred to as capillary zone) has to be taken into account. Physical processes essentially comprise hydraulic and thermal processes in porous media. Heat generation by chemical or biochemical reactions or by radioactive decay is not assumed to be of importance in the case of heat transport in the subsurface. Thus, the relevant processes considered here comprise the flow of water and the heat transport in both the unsaturated and the saturated zones of the subsurface.

2.1.1 Modeling hydraulic processes in porous media

2.1.1.1 Flow in saturated and unsaturated porous media, Darcy's law

Hydraulic processes are important for heat transport whenever advective heat flux, by flowing water, is significant. Also in the case of stagnant or static conditions, the water content plays a role in the thermal parameters. When considering hydraulic processes in connection with thermal propagation, we have to be aware that essential physical parameters of flow processes, like **water density** and **water viscosity**, are temperature dependent. In general, water density ρ_w (kg m^{-3}) and dynamic water viscosity μ_w (Pa s) are considered to be functions of water pressure p_w, concentration c of dissolved substances, and temperature T (°C), or $\rho_w = \rho_w (p_w, c, T)$ and $\mu_w = \mu_w (p_w, c, T)$. While temperature dependence of water density leads to density effects in flow problems (with maximum water density close to 4°C), the temperature dependence of water viscosity leads to decreasing viscosity values for increasing temperature (Figure 2.1). Typical values are shown in Section 2.2. The temperature

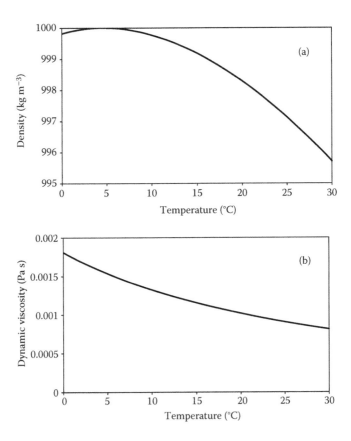

Figure 2.1 Temperature dependence of water density (a) and water viscosity (b). (Data from www.thermexcel.com.)

dependence of water viscosity directly affects **hydraulic conductivity K_w** (m s^{-1}) (Bear 1979), that is,

$$K_w(T) = \frac{kg\rho_w(p_w,c,T)}{\mu_w(p_w,c,T)} \tag{2.1}$$

where **k** (m^2) is the permeability tensor of the porous medium, which depends solely on the geometrical configuration of the porous matrix, and g is the acceleration due to gravity. This means that, for example, hydraulic conductivity at 20°C is about 1.5 times higher than at 5°C due to the change in water viscosity.

A consequence of the temperature dependence is that the **general form of Darcy's law** (Bear 1979) for expressing water fluxes should be used, that is,

$$q_w = -\frac{k}{\mu_w(p_w,c,T)}[\nabla p_w - \rho_w(p_w,c,T)g] \tag{2.2}$$

where q_w (m s^{-1}) is the specific flux vector (Darcy velocity), ∇ is the gradient operator, p_w is the water pressure (Pa) and **g** (m s^{-2}) is the vector of acceleration due to gravity having the form $\mathbf{g} = (0, 0, -g)$, for a vertical z-axis pointing upward. Equation 2.2 can be written as balance of the **forces acting** per unit volume of water:

$$-\frac{q_w\mu_w(p_w,c,T)}{k} - \nabla p_w + \rho_w(p_w,c,T)\mathbf{g} = 0 \tag{2.3}$$

The first term denotes the **friction force**, the second one the **water pressure force**, and the third term is the **gravity force**, all per unit volume.

By introducing a **constant reference temperature** T_0 (e.g., mean temperature at soil surface) and a **constant reference pressure** $p_{w,0}$, and by restricting the analysis to **constant concentrations** $c = c_0$ in the following, the **water density** can be approximated by linear expansion as follows:

$$\rho_w(p_w,T) \simeq \rho_w(p_{w,0},T_0) + \left(\frac{\partial\rho_w}{\partial T}\right)_{T_0,p_{w,0}}(T-T_0) + \left(\frac{\partial\rho_w}{\partial p_w}\right)_{T_0,p_{w,0}}(p_w - p_{w,0})$$

$$\simeq \rho_{w,0}[1 + b_p(p_w - p_{w,0}) + b_T(T - T_0)] \tag{2.4}$$

with $\rho_{w,0} = \rho_w(p_{w,0}, T_0)$. The coefficients b_p and b_T with

$$b_p = \frac{1}{\rho_{w,0}}\frac{\partial\rho_w}{\partial p_w}; \quad b_T = \frac{1}{\rho_{w,0}}\frac{\partial\rho_w}{\partial T} \tag{2.5}$$

are the compressibility and the thermal volume expansion coefficients for water. In an alternative formulation for ρ_w (p_w, T), the compressibility of water is neglected:

$$\rho_w(p_w,T) = \rho_w(p_w,T_0) + \Delta\rho_w(T, T_0) \simeq \rho_w(p_w, T_0) \, [1 + b_T \, (T - T_0)] \quad (2.6)$$

Based on Equation 2.6, Darcy's law (Equation 2.2) can be reformulated as follows:

$$q_w = -\frac{k}{\mu_w(p_w,T)}\left[\nabla p_w - \left(\rho_w(p_w,T_0) + \Delta\rho_w(T,T_0)\right)g\right] \quad (2.7)$$

and the balance Equation 2.3 of acting forces is given by

$$-\frac{q_w\mu_w(p_w,T)}{k} - \nabla p_w + \rho_w(p_w,T_0)g + \Delta\rho_w(T,T_0)g = 0 \quad (2.8)$$

Based on this formulation, the effect of density variations due to temperature changes may be assessed. The ratio between the buoyancy force $\Delta\rho_w g$ acting vertically and the horizontal friction force from Darcy's law in a regional flow is in absolute values:

$$G = \frac{\Delta\rho_w}{\rho_{w,0}}\frac{k\mu_w}{q} = \frac{\Delta\rho_w}{\rho_{w,0}}\frac{K_w}{q} = \frac{\Delta\rho_w}{\rho_{w,0}I_{hor}} \quad (2.9)$$

The symbol I_{hor} is the horizontal flow gradient. Oostrom et al. (1992) called G a **stability number,** which they obtained from dimensional analysis. Based on experimental investigation, they determined a critical stability number of about $G_c = 0.3$. Woumeni and Vauclin (2006) confirmed the usefulness of the ratio in their field study on the coupled effects of aquifer stratification, fluid density, and groundwater fluctuations on dispersivity in solute transport. The stability number can be used to **assess the importance of density effects.** For **example,** if we inject water at a temperature of $T = 9°C$ into an aquifer of initial temperature $T_0 = 12°C$, and with a horizontal flow gradient of $I_{hor} = 0.001$, the density difference $\Delta\rho_w$ is about 0.28 kg m^{-3}, and the ratio of Equation 2.9 is 0.28, thus only slightly smaller than the critical value. Even if density effects are present, due to mixing effects, these effects may gradually be reduced away from the injection point. Neglecting density effects may be acceptable at substantial temperature differences between injected water and groundwater. This was shown by Ma and Zheng (2010) based on their modeling study of the Hanford site (United States). They found that for thermal tracer experiments, model errors due to ignoring density effects are insignificant for temperature differences as large as 15°C across the entire model domain. Ward et al. (2007) concluded from their theoretical and numerical analysis that the relevance of density effects in aquifer storage and recovery depends on the relative influences of density difference, hydraulic conductivity, pumping rates, injected radius, storage duration, and dispersivity, thus confirming the importance of the flow condition.

In the case of thermal use of shallow subsurface systems with restricted temperature changes, **water density and viscosity are often used as constants in thermohydraulic models**, and thus, temperature dependence is disregarded. As a consequence, water flow is not affected by heat transport, and both equations can be handled and solved in an uncoupled and sequential manner. Since **flow processes represent an important element in thermal processes**, they are briefly compiled here.

Hydraulic processes can take place in both the **saturated and unsaturated zones**. For convenience, both zones are treated here simultaneously. In unconfined aquifers, the upper limit of the groundwater zone consists of the water table. The level of the water table is usually defined as the location with atmospheric water pressure (zero relative water pressure). Consequently, the domain above the water table is the unsaturated zone (or capillary zone, Figure 2.2). In principle, the capillary zone is hydraulically unsaturated by the simultaneous presence of water and air in a control volume. However, part of the capillary zone close to the water table can still be hydraulically saturated (saturated or quasi-saturated capillary fringe; Figure 2.2). The corresponding related relative water pressure in the capillary zone is negative (suction).

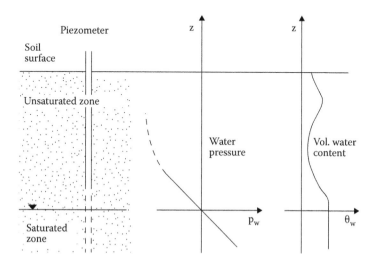

Figure 2.2 Capillary zone above a water table; vertical water pressure profile $p_w(z)$ and vertical profile of the volumetric water content $\theta_w(z)$ (schematic).

Flow in saturated porous media with constant water density is usually described by the following form of **Darcy's law**, which is directly based on Darcy's postulation:

$$\mathbf{q}_w = -\mathbf{K}_w \, \nabla h_w \tag{2.10}$$

where h_w (m) is the piezometric head (or head) with

$$h_w = z + \frac{p_w}{\rho_w g} \tag{2.11}$$

Darcy's law states that the flow rate is proportional to the head gradient. The variable z(m) is the vertical coordinate (positive upward). The equations can also be obtained from Equations 2.1 and 2.2 by setting the water density ρ_w to a constant.

Hydraulic conductivity $\mathbf{K}_w(\mathbf{x})$ at location $\mathbf{x}(x, y, z)$ (m) usually exhibits strong spatial variability due to nonhomogeneity of porous media, for example, formations with layers, lenses, etc. (e.g., in Jussel et al. 1994; Bayer et al. 2011). Moreover, it may show a directional behavior thus causing **anisotropic conditions**. Such conditions prevail, for example, in layered porous media where the largest hydraulic conductivity is parallel to the layering and the minimum hydraulic conductivity perpendicular to it (Figure 2.3). In general, formulation of the hydraulic conductivity coefficient is a symmetric second rank tensor (Bear 1979). Due to its symmetry, the number of components is six. For isotropic porous media, the tensor reduces to a single scalar quantity K_w. For anisotropic conditions, any symmetric tensor can be transformed into a diagonal matrix by rotation of the coordinate system by horizontal and vertical angles to the so-called **principal axes** x', y', and z', leading to the corresponding Darcy law with diagonal tensor:

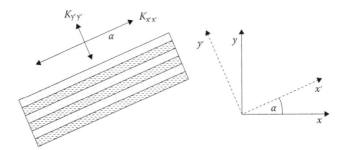

Figure 2.3   Inclined layer consisting of coarse (dotted) and fine (dashed) layers, leading to anisotropy in hydraulic conductivity of the layer as a whole; related coordinate systems with principal directions x' (parallel to the layering) and y' (perpendicular to layering).

$$q = \begin{Bmatrix} q_x \\ q_y \\ q_z \end{Bmatrix} = -\mathbf{K}_w \nabla h_w = - \begin{bmatrix} K_{xx} & K_{xy} & K_{xz} \\ K_{yx} & K_{yy} & K_{yz} \\ K_{zx} & K_{zy} & K_{zz} \end{bmatrix} \begin{Bmatrix} \dfrac{\partial h}{\partial x} \\ \dfrac{\partial h}{\partial y} \\ \dfrac{\partial h}{\partial z} \end{Bmatrix} = - \begin{bmatrix} K_{x'x'} & 0 & 0 \\ 0 & K_{y'y'} & 0 \\ 0 & 0 & K_{z'z'} \end{bmatrix} \begin{Bmatrix} \dfrac{\partial h}{\partial x'} \\ \dfrac{\partial h}{\partial y'} \\ \dfrac{\partial h}{\partial z'} \end{Bmatrix}$$

$$(2.12)$$

Therefore, the number of essential components is reduced to three. In layered systems, it is often assumed that hydraulic conductivity that is parallel to layering is isotropic. Such a case can be described by one value parallel (maximum value) and one perpendicular to the layering (minimum value), one vertical angle of the layer plane with respect to the horizontal plane, and one horizontal angle defining the position of the plane. Moreover, very often, horizontal layering is observed in aquifers, thus reducing the hydraulic conductivity components to two, the horizontal and vertical hydraulic conductivities, that is, K_{hor}, and K_{vert}.

Unsaturated porous media are, in general, characterized by the presence of a continuous air phase besides a continuous water phase (Figure 2.2). Flow can take place in both fluid phases. The spatial distribution of both phases strongly depends on the wetting properties with respect to the solid phase (rock material). The volumetric fraction of both phases can be time dependent. The volumetric fraction of the water phase in a control volume is called **volumetric water content** θ_w (water volume per unit volume of porous medium). In unsaturated porous media, the volumetric water content θ_w is smaller than the **porosity** ϕ (interconnected pore volume per unit volume of porous medium). In a simplified manner, the water flux equation (Darcy's law) can be generalized from saturated flow conditions by analogy and adjusted accordingly. Consider the following situation. A vertical column with a homogeneous porous medium is recharged by a uniform steady-state infiltration rate at the top inflow face. If the column is long enough, practically uniform flow will be established with constant water content and constant (negative relative) water pressure, provided the infiltration rate is smaller than the hydraulic conductivity of the porous medium at saturation. Such a water flow system is characterized by two kinds of boundaries within the microscopic flow system. On the one hand, the solid phase represents the boundary as in saturated flow. On the other hand, the air–water interface is a boundary. The difference to saturated conditions lies in the fact that $\theta_w < \phi$. Furthermore, the resistance at the water–air interface is different. Even so, it can be presumed that **Darcy's law** for the water phase can be generalized for **constant water density** as follows (e.g., Bear 1979):

$$q_w = -K_w(S_w)\nabla\left(z + \frac{p_w}{\rho_w g}\right) \tag{2.13}$$

where S_w is the **water saturation**, with $S_w = \theta_w/\phi$. Still, it is formally equivalent to Equation 2.10. Similarly, Darcy's law for the air phase can be formulated correspondingly (Bear 1979). Moreover, it can be generalized according to Equation 2.2 in order to include density effects. It can be expected that **hydraulic conductivity strongly depends on the water saturation**, that is $K_w(S_w)$, which is a phenomenological relationship with strongly decreasing hydraulic conductivity values for decreasing water saturation. Frequently used **models** for $K_w(S_w)$ are those of Brooks and Corey (1966) and van Genuchten (1980). Brooks and Corey's model states that

$$K_w(S_w) = K_{w,sat}\left(\frac{S_w - S_{w,r}}{1 - S_{w,r}}\right)^{3 + 2/\lambda_{BC}} \tag{2.14}$$

where $S_{w,r}$ is the residual saturation (water saturation, which cannot be drained by gravitational effects only), $K_{w,sat} = K(S_w = 1)$, and λ_{BC} is the pore distribution index. Typical values in granular porous media are $S_{w,r} = 0.1$ and $\lambda_{BC} = 2$. Van Genuchten's (1980) model for $K_w(S_w)$ is

$$K_w(S_w) = K_{w,sat}S_{w,e}^{1/2}\left[1 - \left(1 - S_{w,e}^{1/m_{VG}}\right)^{m_{VG}}\right]^2 \tag{2.15}$$

With the parameters m_{VG}, n_{VG}, and $S_{w,r}$, and the effective water saturation $S_{w,e}$:

$$S_{w,e} = \frac{S_w - S_{w,r}}{1 - S_{w,r}} \tag{2.16}$$

Usually, $m_{VG} = 1 - 1/n_{VG}$. A modified formulation for improved description near saturation can be found in Schaap and van Genuchten (2006).

It is usually assumed in models that the relation $K_w(S_w)$ is identical for static, steady-state, and transient conditions. Furthermore, hysteresis effects in $K_w(S_w)$ are small and are often neglected. Therefore, the relation is considered unique.

2.1.1.2 Water mass balance, volume balance, flow equation

The general **mass balance** for the water phase in a **saturated porous medium** can be formulated for a unit control volume of the porous medium (Figure 2.4):

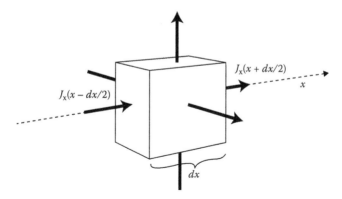

Figure 2.4 Unit control volume of a porous medium with flux components through horizontal fluxes in x-direction (schematic).

$$\frac{\partial(\phi\rho_w)}{\partial t} = -\nabla\cdot(\rho_w\mathbf{q}_w) + w\rho_w \tag{2.17}$$

where $\nabla\cdot()$ is the divergence operator, t is the time, and w is a hydraulic source/sink term (injection/extraction of water volume per unit volume of porous medium per unit time) (s^{-1}). Note that porosity $\phi(p_w(t))$ generally depends on water pressure and water density, while water density $\rho_w(p_w(t), T(t))$ depends on water pressure and temperature. The equation states that the rate of change of water mass over time equals water inflow minus outflow for a unit control volume. Assuming linear dependence of porosity from water pressure, that is, $\phi(p_w)$, according to

$$\phi(p_w) \simeq \phi(p_{w,0}) + a\,(p_w - p_{w,0}) \tag{2.18}$$

and by inserting in Equation 2.4, we obtain the following **water mass balance equation**:

$$\frac{\partial(\phi\rho_w)}{\partial t} = \phi\frac{\partial\rho_w}{\partial t} + \rho_w\frac{\partial\phi}{\partial t} = \left[(\rho_w(T)a + \rho_{w,0}\phi b_p)\frac{\partial p_w}{\partial t} + \rho_{w,0}\phi b_T\frac{\partial T}{\partial t}\right] \tag{2.19}$$

We may use the classical concept of **specific storativity** S_s (m^{-1}) with

$$S_s = (\rho_w a + \rho_{w,0}\,\phi b_p)g \tag{2.20}$$

and insert Darcy's law to obtain

$$\frac{S_s}{\rho_{w,0}g}\frac{\partial p_w}{\partial t}+\phi b_T\frac{\partial T}{\partial t}$$

$$=\nabla\cdot\left[\mathbf{K}_w\left(\frac{\nabla p_w}{g\rho_{w,0}}+\nabla z+b_p\nabla z(p_w-p_{w,0})+b_T\nabla z(T-T_0)\right)\right]+\frac{\rho_w(T)}{\rho_{w,0}}w$$

$$(2.21)$$

As an **alternative**, we can insert Equation 2.6 for the water density $\rho_w(T)$ and **Hubbert's** definition of a **water pressure–dependent piezometric head** $h_w(p_w)$, applied for isothermal conditions ($T = T_0$), with

$$h_w(p_w,T_0)=z+\int_{p_{w,0}}^{p_w}\frac{1}{\rho_w(p_w,T_0)g}\,dp_w \qquad (2.22)$$

In the context of temperature-dependent flow, it can be considered as an **equivalent head**. With this formulation, the pressure gradient has the form

$$\nabla p_w = \rho_w(p_w,T_0)g\nabla h_w - \rho_w(p_w,T_0)g\nabla z \qquad (2.23)$$

and the **water balance equation** is

$$S_s\frac{\partial h_w}{\partial t}+\phi b_T\frac{\partial T}{\partial t}=\nabla\cdot\left[\mathbf{K}_w(\nabla h_w+b_T\nabla z(T-T_0))\right]+\frac{\rho_w(T)}{\rho_{w,0}}w \qquad (2.24)$$

This form was also presented by Mercer et al. (1982) and Clauser (2003). However, both start from the **equivalent piezometric head** h_w according to

$$h_w=z+\frac{p_w}{\rho_{w,0}g} \qquad (2.25)$$

Molson et al. (1992) start from a simplified version for their model neglecting the second term in Equation 2.24 and obtain the water balance equation

$$S_s\frac{\partial h_w}{\partial t}=\nabla\cdot\left[\mathbf{K}(\nabla h_w+\rho_{rel}(T)\nabla z)\right]+w \qquad (2.26)$$

with the **relative water density** $\rho_{rel}(T)$ according to

$$\rho_{rel}(T)=\frac{\rho_w(T)}{\rho_{w,0}}-1 \qquad (2.27)$$

The mass **balance for the water phase in unsaturated porous media with constant water density** ρ_w can be formulated as volume balance:

$$\phi \frac{\partial S_w}{\partial t} = -\nabla \cdot \mathbf{q}_w + w \qquad (2.28)$$

The above formulation does not take into account compressibility effects of water and porous matrix, nor does it consider phase exchange processes between the phases, due to, for example, evaporation or condensation of water or air dissolution in water. Similarly, the mass balance for the air phase can be postulated correspondingly (Bear 1979). For saturated conditions, that is, $S_w = 1$, the transient term vanishes. By inserting Darcy's law into the volume balance equations and by **neglecting the influence of the air phase on water flow**, a nonlinear differential equation of second order, the **unsaturated flow equation**, also known as **Richards' equation**, is obtained:

$$\phi \frac{\partial S_w}{\partial t} = \nabla \cdot \left(\mathbf{K}_w (S_w) \nabla \left(z + \frac{p_w}{\rho_w g} \right) \right) + w \qquad (2.29)$$

Neglecting air flow in the unsaturated zone means that air pressure is taken as atmospheric in the soil, with **zero relative air pressure**, that is, $p_a = 0$. Equation 2.29 contains the two variables S_w and p_w. Therefore, a further relation, the relation between water saturation S_w, the water pressure p_w, and the air pressure p_a, is required in order to solve the problem. The relation is known as the **water retention curve** $S_w(p_c)$, where p_c is called **capillary pressure**, with $p_c = p_a - p_w$ in general, or $p_c = -p_w$ when air flow is neglected (with $p_a = 0$). The water retention curve is a phenomenological relationship, with decreasing water saturation for decreasing water pressure values. Frequently used **models** for $S_w(p_c)$ are those of Brooks und Corey (1966) and van Genuchten (1980).

Brooks and Corey's model states that

$$S_{w,e} = \frac{S_w - S_{w,r}}{1 - S_{w,r}} = \left(\frac{p_b}{p_c} \right)^{\lambda_{BC}} ; \quad p_c \geq p_b$$
$$S_{w,e} = 1; \qquad\qquad\qquad 0 \leq p_c \leq p_b \qquad (2.30)$$

where p_b is the air entry pressure (suction at which air enters the porous medium in drainage processes). The parameter λ_{BC} is the same as that used in Equation 2.14.

Van Genuchten's (1980) model for $S_{w,e}$ is

$$S_{w,e} = \frac{S_w - S_{w,r}}{1 - S_{w,r}} = \left(\frac{1}{\left(\left(1 + \left| \alpha_{VG} \frac{p_c}{\rho_w g} \right|^{n_{VG}} \right)^{m_{VG}} \right)} \right) \tag{2.31}$$

with the parameters m_{VG}, n_{VG}, and α_{VG}.

It is usually assumed in models that the relation $S_w(p_c)$ is identical for static, steady-state, and transient conditions. **Hysteresis effects** in $S_w(p_c)$ are present but are often neglected. Therefore, it is assumed that the relation is unique.

For **saturated conditions**, the flow Equation 2.17 for constant water density ρ_w but a compressible porous matrix is

$$S_s \frac{\partial h_w}{\partial t} = \nabla \cdot (\mathbf{K}_w (S_w) \nabla h_w) + w \tag{2.32}$$

The **specific storativity** S_s of the porous medium can be interpreted as water volume change in a unit control volume per unit change of the piezometric head h_w. For a more general discussion of saturated and unsaturated flow models, the reader is referred to Bear and Cheng (2010).

For **freezing soils and aquifers**, Equation 2.29 can be extended (Williams and Smith 1989; Hansson et al. 2004) for saturated and unsaturated conditions as follows:

$$\frac{\partial \theta_w}{\partial t} + \frac{\rho_i}{\rho_w} \frac{\partial \theta_i(T)}{\partial t} = \phi \frac{\partial S_w}{\partial t} + \frac{\rho_i}{\rho_w} \phi \frac{\partial S_i(T)}{\partial t} = \nabla \cdot \left[\mathbf{K}_w (p_w, T) \nabla \left(z + \frac{p_w}{\rho_w g} \right) \right] \tag{2.33}$$

where the subscript $_{-w}$ means liquid water, and θ_i and S_i are the volumetric ice content and the ice saturation. The second term on the left-hand side of Equation 2.33 expresses the rate of change of ice mass, measured as equivalent water volume, per unit time and unit volume. Flow (right-hand side) only occurs for liquid water. For the saturation degrees, the overall condition has to be fulfilled:

$$S_w + S_a + S_i = 1 \tag{2.34}$$

Equation 2.33 has to be coupled with the corresponding equation for heat transport, Equation 2.89.

2.1.1.3 Initial and boundary conditions

The **initial condition** for flow problems consists of specifying water pressure $p_w(\mathbf{x}, t = 0)$ in general, or piezometric head $h_w(\mathbf{x}, t = 0)$ for saturated zone models. The symbol $\mathbf{x}(x, y, z)$ denotes an arbitrary location vector.

Boundary conditions for flow problems are specified values of the variable at a boundary section B_1, specified water flux through a boundary section B_2, or flux through a semipermeable boundary section B_3.

Specified values p_{w1} or h_{w1} at a boundary section B_1 (Figure 2.5a) are, according to a first type or Dirichlet boundary condition:

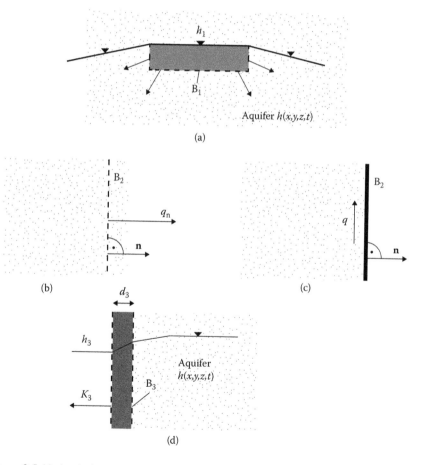

Figure 2.5 Hydraulic boundary conditions (schematic). (a) Surface water body with prescribed head, B_1; (b) prescribed flux into aquifer, B_2; (c) impermeable boundary, B_2; (d) semipermeable boundary, B_3.

$$p_w(\mathbf{x}_{B_1}, t) = p_{w_1}(\mathbf{x}_{B_1}, t); \quad \in B_1$$
$$h_w(\mathbf{x}_{B_1}, t) = h_{w_1}(\mathbf{x}_{B_1}, t); \quad \in B_1 \tag{2.35}$$

An example for B_1 is the direct connection of an aquifer with a surface water body at water level h_{w1}, thus controlling the water pressure in the adjacent aquifer.

Specified water flux $q_{w2,n}$ (normal component, positive in the direction of n) through a boundary section B_2 (Figure 2.5b), according to a second type or Neumann boundary condition, requires

$$q_{w,n}(\mathbf{x}_{B_2}, t) = q_{w2,n}(\mathbf{x}_{B_2}, t) = \left(-K_{w,n} \frac{\partial h_w}{\partial n} \right)_{\mathbf{x}_{B_2}, t} \quad \in B_2 \tag{2.36}$$

An example for B_2 is a given infiltration rate through the soil surface, which eventually leads to recharge of the aquifer. Prescribed lateral inflow into an aquifer, for example, from hill slopes, or from an upstream valley, or prescribed inflow or extraction rates in boreholes and wells are some other examples. For an **impermeable boundary** (Figure 2.5c), $q_n = 0$.

The flux condition for a **semipermeable boundary** section B_3 (Figure 2.5d) in an isotropic aquifer is

$$-K_w \frac{\partial h_w}{\partial n}(\mathbf{x}_{B_3}, t) = K_3 \frac{\left[h_w(\mathbf{x}_{B_3}, t) - h_3 \right]}{d_3} = l_3 \cdot \left[h_w(\mathbf{x}_{B_3}, t) - h_3 \right]; \quad \in B_3 \tag{2.37}$$

The symbol l_3 denotes the **leakage coefficient**. Note that for positive head gradients, the normal flux is directed from the solution domain to the outside region. The condition has the form of a mixed type or Cauchy boundary condition. An example for B_3 is a semipermeable river bottom, which exhibits reduced hydraulic conductivity values due to clogging effects caused by, for example, the deposition of fine sediments on the river bottom. The related additional resistance is accounted for by the leakage coefficient.

2.1.1.4 Two-dimensional flow models for saturated regional water flow

Regional water flow in shallow aquifers is frequently described by vertically averaged, two-dimensional horizontal flow models. This type of model is restricted to saturated conditions, where the aquifer thickness is small compared to the lateral extent. Consequently, vertical flow components are disregarded. The vertical integration of the flow equation for horizontally

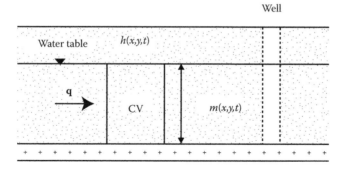

Figure 2.6 Schematic cross section through a shallow, extended, and unconfined aquifer, with control volume (CV) extending from aquifer bottom to the water table.

isotropic aquifers leads to the following **volume balance equation** for a control volume, based on Equation 2.32 (see Figure 2.6):

$$S\frac{\partial h_{\mathrm{w}}}{\partial t} = \nabla \cdot (\mathbf{K}_{\mathrm{w}} m \nabla h_{\mathrm{w}}) + W \tag{2.38}$$

where S is the aquifer **storativity** (water volume change per unit horizontal area per unit change of piezometric head h_{w}). For confined aquifers, the storativity $S = S_{\mathrm{s}}\, m$ is a small value, considering compressible deformation of water and porous matrix. For unconfined conditions, the coefficient is called **specific yield** or drainable porosity (water volume change per unit horizontal area per unit rise/decline of water table; usually smaller than porosity ϕ). The specific yield is usually much larger than the storativity of confined aquifers. The term $(\mathbf{K}_{\mathrm{w}}\, m)$ is called aquifer **transmissivity** (in general, a tensor), where m (m) is the **aquifer thickness**, which is, for unconfined aquifers

$$m(\mathbf{x}, t) = h_{\mathrm{w}}(\mathbf{x}, t) - z_{\mathrm{bot}}(\mathbf{x}) \tag{2.39}$$

The symbol $z_{\mathrm{bot}}(\mathbf{x})$ denotes the **aquifer bottom elevation**. $W(\mathbf{x}, t)$ represents the source/sink term (water volume injection/extraction per unit area and per unit time) (m s^{-1}). This can be a **recharge term** $N(\mathbf{x}, t)$, or a local **source/sink**, such as a pumping well (discharge rate per unit area, Q/A). The aquifer thickness m is a function of the unknown flow variable $h_{\mathrm{w}}(\mathbf{x},t)$. Equation 2.38 is therefore **nonlinear for unconfined aquifers**. For **confined aquifers**, the aquifer thickness is

$$m(\mathbf{x}) = z_{\mathrm{top}}(\mathbf{x}) - z_{\mathrm{bot}}(\mathbf{x}) \tag{2.40}$$

where z_{top} is the top of the aquifer. In this case, the volume balance equation is linear.

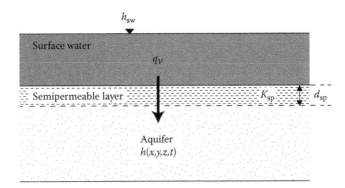

Figure 2.7 Semipermeable layer between surface water and aquifer (schematic cross section).

Interaction with surface water can be modeled by the local recharge term $N(\mathbf{x}_{sw})$ (m s^{-1}) for the horizontal area of the surface water body. In the case of a semipermeable layer (e.g., leaky rivers; Figure 2.7), this term is usually approximated by using Darcy's law for the vertical flux q_{vert}:

$$N_{sw} = q_{vert} = K_{sp} \frac{h_{sw} - h}{d_{sp}} = l(h_{sw} - h) \tag{2.41}$$

where K_{sp} is the hydraulic conductivity of the semipermeable layer, h_{sw} is the head of the surface water, d_{sp} is the thickness of the semipermeable layer, and l is the **leakage coefficient**. In the case where the groundwater table is below the bottom of the surface water, the vertical flux is usually approximated by

$$N_{sw} = q_{vert} = K_{sp} \frac{h_{sw} - z_{bot}}{d_{sp}} = l(h_{sw} - z_{bot}) \tag{2.42}$$

neglecting the water suction below the semipermeable layer. Note that in this case, a first-order term in $h_w(\mathbf{x})$ shows up in the flow Equation 2.38.

Initial and boundary conditions are formulated in a manner corresponding to the three-dimensional case.

2.1.2 Modeling thermal processes in porous media

Thermal processes in the subsurface essentially comprise heat conduction in solid materials (rock material, grains), in soil water, and in soil air, as well as heat advection in flowing water including thermal dispersion effects. In principle, they also include phase change processes (evaporation, condensation, freezing, thawing), which are related to considerable energy transfers. However, for the thermal use of shallow subsurface systems, they are not usually significant and are, therefore, often disregarded. If, nevertheless, phase change processes have to be taken into account for a particular problem, the reader is referred to the literature of, for example, Williams and Smith (1989) for freezing soils. An introduction to heat transfer in fluids and solid materials can be found, for example, in Incropera et al. (2007). In these contexts, we will restrict ourselves to a brief introduction only.

2.1.2.1 Heat storage, heat capacity, and advective heat transport

The **thermal energy** ΔE (J) stored in a water volume V_w (m^3) at temperature T (°C), with respect to a reference temperature T_0, with $\Delta T = T - T_0$, can be expressed as follows:

$$\Delta E = V_w c_w \rho_w \Delta T \tag{2.43}$$

The coefficient c_w is the **specific heat capacity** (or specific heat) of water, which is related to a unit mass of water (J kg^{-1} K^{-1}). More precisely, it is the specific heat, which is usually determined for constant pressure conditions. The **volumetric heat capacity of water** C_w (J m^{-3} K^{-1}) related to a unit volume of water is

$$C_w = \rho_w c_w \tag{2.44}$$

The **reference temperature** T_0 can be, for example, the initial temperature or the mean annual temperature, or 0°C, depending on the situation to be investigated.

For **solid materials**, the specific heat capacity is, correspondingly, c_s, and the volumetric heat capacity C_s. In a similar manner, we get the specific heat capacity c_a and the volumetric heat capacity C_a for **air**.

For **saturated aquifer material** with porosity ϕ, the **volumetric heat capacity** C_m (J m^{-3} K^{-1}) is usually expressed by the **weighted arithmetic average** of the (mixed) values for water and solid material, weighted by the corresponding volumetric fractions:

$$C_m = \phi C_w + (1 - \phi) C_s \tag{2.45}$$

It represents the heat storage of water and solid matrix, presuming that the mean temperature for the water and the solid phase within a control volume are equal. For unsaturated conditions, with water, solid, and air as well as vapor phase, the volumetric heat capacity is, according to Whitaker (1977),

$$C_m = \phi S_w C_w + \phi S_a (C_a + C_v) + \phi S_i C_i + (1 - \phi) C_s \qquad (2.46)$$

where $S_a = 1 - S_w - S_i$ is the air saturation, and C_a, C_v, and C_i are the volumetric heat capacities of air, vapor, and ice, respectively. However, the contribution of air and vapor is normally small and, therefore, often neglected. Values of volumetric heat capacities of water, sand packings, and aquifers are presented in Section 2.2.

Based on the storage concept, the **advective heat flux** J_w (J s^{-1}) for water flowing at the discharge rate Q (m^{3}s^{-1}) and temperature T with respect to a reference temperature T_0 through a cross section can be expressed as

$$J_w = Q C_w (T - T_0) \qquad (2.47)$$

Equation 2.47 applies for saturated and unsaturated conditions. The corresponding **specific advective heat flux** vector j_w (heat flux through unit area) (J m^{-2} s^{-1}) caused by the specific water flux vector $\mathbf{q}$ (Darcy flux) (m s^{-1}) is therefore

$$j_w = q C_w (T - T_0) \qquad (2.48)$$

2.1.2.2 Heat conduction

The **heat flux** J_w **by heat conduction** (or thermal diffusion) through a cross section of water with area A in direction x normal to the area A is proportional to A and to the temperature gradient $\Delta T / \Delta x$:

$$J_{w,x} = A \lambda_w \frac{\Delta T}{\Delta x} \qquad (2.49)$$

ΔT is the temperature increment over the spatial increment Δx. The coefficient λ_w is the **thermal conductivity of water** (W m^{-1} K^{-1}). Equation 2.49 is known as **Fourier's law**. The corresponding **specific heat flux vector** j_w, assuming isotropic thermal conductivity, is

$$j_w = -\lambda_w \nabla T \qquad (2.50)$$

The specific heat flux j_s through a cross section of **solid material** is expressed in a similar manner, according to Fourier, as

$$j_s = -\lambda_s \nabla T \qquad (2.51)$$

using the thermal conductivity λ_s of solid material (rock material). For **aquifer material** consisting of water, rock, and air, the thermal conductivity λ_m

is an **equivalent parameter**, which is a function of the composition of solid material, water, and air. The corresponding heat flux vector is

$$j_m = -\lambda_m \nabla T \tag{2.52}$$

Various techniques exist to express the thermal conductivity λ_m of saturated and unsaturated porous media, given values for water, ice, air, and rock material.

The simplest way to calculate the thermal conductivity consists of the **weighted arithmetic average** of thermal conductivities, according to Whitaker (1977):

$$\lambda_m = \phi S_w \lambda_w + \phi S_a \lambda_a + \phi S_i \lambda_i + (1 - \phi)\, \lambda_s \tag{2.53}$$

Again, the influence of the air phase is small. Usually, this value overpredicts the effective thermal conductivity of an aquifer. For **saturated porous media**, it is well known (see, e.g., Dagan 1989) that the **weighted arithmetic average** is the **upper bound** for the effective thermal conductivity (for $S_i = 0$):

$$\lambda_{m,\,sat} = \phi \lambda_w + (1 - \phi)\, \lambda_s \tag{2.54}$$

On the other hand, the **lower bound** is represented by the **weighted harmonic average**:

$$\lambda_{m,sat} = \frac{1}{\left[\dfrac{\phi}{\lambda_w} + \dfrac{(1-\phi)}{\lambda_s} \right]} \tag{2.55}$$

Dagan (1989) proposes the so-called **self-consistent approximation** (also termed renormalization approximation) to express thermal conductivity of saturated granular porous media ($S_w = 1$). The **effective thermal conductivity** is then

$$\lambda_{m,sat} = \frac{-b + \sqrt{b^2 + 8\lambda_w \lambda_s}}{4} \tag{2.56}$$

with the coefficient b

$$b = \lambda_w\,(3\phi - 1) - \lambda_s\,(2 - 3\phi) \tag{2.57}$$

Kunii and Smith (1960) derived equations to estimate the effective thermal conductivity $\lambda_{m,\,sat}$ of unconsolidated particle beds with uniform grain size distribution. For a porous medium saturated with water, $\lambda_{m,\,sat}$ is calculated as

$$\lambda_{m,sat} = \lambda_w \left[\phi + \frac{(1-\phi)}{\varepsilon + \dfrac{2}{3} \dfrac{\lambda_w}{\lambda_s}} \right] \tag{2.58}$$

The function ε depends on the effective thickness of water films adjacent to the contact surface of two solid particles and the number of contact points n of a particle with neighbors. It is determined by

$$\varepsilon = \frac{1}{3\kappa} \cdot \frac{\left(\dfrac{\kappa-1}{\kappa}\right)^2 \sin^2 \alpha}{\ln(\kappa - (\kappa-1)\cos\alpha) - \dfrac{\kappa-1}{\kappa}(1-\cos\alpha)} \tag{2.59}$$

The symbol κ denotes $\kappa = \lambda_s/\lambda_w$. The angle α is determined by $\sin^2\alpha = 1/n$. The function ε is evaluated for both a loose packing ($\phi_1 = 0.476$) and the densest packing with uniform grains ($\phi_2 = 0.260$). The values for the corresponding number of contact points are $n_1 = 1.42$ and $n_2 = 6.93$. For arbitrary porosity, it is suggested that the function ε is determined by linear interpolation:

$$\begin{aligned}
\varepsilon &= \varepsilon_2 \quad \text{for } \phi \leq \phi_2; \\
\varepsilon &= \varepsilon_2 + (\varepsilon_1 - \varepsilon_2)\frac{\phi - \phi_2}{\phi_1 - \phi_2} \quad \text{for } \phi_2 \leq \phi \leq \phi_1; \\
\varepsilon &= \varepsilon_1 \quad \text{for } \phi \geq \phi_1
\end{aligned} \tag{2.60}$$

de Vries (1963) proposed a method to calculate thermal conductivity of soils using volume fraction and physical properties of its constituents. His theory is based on Maxwell's approach and Burger's extension (Woodside and Messmer 1961) to calculate electrical conductivity of two-phase materials. For the extension, ellipsoidal soil particles with the axes a, b, and c are assumed, which are not in contact with each other and are embedded in a continuous medium of water. The effective **thermal conductivity** λ_m of a soil system consisting of n solid components is calculated as follows:

$$\lambda_{\mathrm{m}} = \frac{\theta_{\mathrm{w}}\lambda_{\mathrm{w}} + \displaystyle\sum_{i=1}^{n} k_i\theta_i\lambda_i}{\theta_{\mathrm{w}} + \displaystyle\sum_{i=1}^{n} k_i\lambda_i} \qquad (2.61)$$

where $i = 1$, n are soil components (mineral or organic soil particles), and θ_i is the volumetric fraction (with respect to unit soil volume). The weighting factors k_i are estimated from the shape of the particles and the thermal conductivities of water and the soil constituents:

$$k_i = \frac{1}{3}\left[\frac{1}{1+\left(\dfrac{\lambda_i}{\lambda_{\mathrm{w}}}-1\right)g_a} + \frac{1}{1+\left(\dfrac{\lambda_i}{\lambda_{\mathrm{w}}}-1\right)g_b} + \frac{1}{1+\left(\dfrac{\lambda_i}{\lambda_{\mathrm{w}}}-1\right)g_c}\right] \qquad (2.62)$$

where g_a is the shape factor of the ellipsoid in the a-direction. For spherical grains, the shape factors are $g_a = g_b = g_c = 1/3$. In this form, the equation corresponds to Maxwell's equation. The quantity k_i is conceived as the ratio of the average temperature gradients in the particles of types i and the corresponding quantity in the water. For the shape factors, de Vries (1963) used the theory of the dielectric constant as analogy. In general, the shape function g_a is found by the integral

$$g_a = \frac{abc}{2}\int_0^\infty \frac{du}{(a^2+u)^{3/2}(b^2+u)^{1/2}(c^2+u)^{1/2}} \qquad (2.63)$$

The shape functions g_b and g_c are obtained in a similar manner, fulfilling the condition that $g_a + g_b + g_c = 1$. Unsaturated conditions can be considered in Equation 2.61 by treating the air as solid particles with the corresponding θ_a and λ_a. Woodside and Messmer (1961), Farouki (1981), Giakoumakis (1994), and Tarnawski and Wagner (1993) used the de Vries model successfully for their studies of unsaturated soils. However, it has to be noted that the de Vries model does not take into account nonuniform grain size distributions. Very often it is assumed that $g_a = g_b$. The value for g_c is $1-2g_a$. This leaves one value undetermined, which is often utilized as a fitting factor. Campbell et al. (1994) extended the de Vries theory and considered both the water and air phases as continuous functions by introducing a fluid thermal conductivity of the form

$$\lambda_f = \lambda_a + f_w (\lambda_w - \lambda_a) \tag{2.64}$$

where f_w is an empirical weighting function depending on the soil, which ranges from 0, in dry soils, to 1 in saturated soils.

In their laboratory experiments with unconsolidated packs of quartz grains, glass beads, and lead shot, **Woodside and Messmer** (1961) found that the formula of Maxwell underestimated the effective thermal conductivity. Both the approach of de Vries (1963) and of Kunii and Smith (1960) gave fair agreement with their observed data. This holds true also for the weighted geometrical mean model for the essential range of $\lambda_s/\lambda_w \le 20$, which was also demonstrated by Menberg et al. (2013a) using various sands and silts from a well in Southern Germany.

Based on **Johansen's** (1975) model, **Farouki** (1981) suggested the following procedure. Thermal conductivity λ_m for porous media consisting of dry crushed rock material ($S_w = 0$) is empirically described by

$$\lambda_{m,\,dry} = (0.039\phi)^{-2.2} \tag{2.65}$$

and for natural soils

$$\lambda_{m,dry} = \frac{0.137\rho_b + 64.7}{2700 - 0.947\rho_b} \tag{2.66}$$

where ρ_b is the dry density of the packing. For saturated conditions ($S_w = 1$), the thermal conductivity λ_m of porous media is approximated by the **weighted geometric average** (Woodside and Messmer 1961) as follows:

$$\lambda_{m,sat} = \lambda_w^\phi \lambda_s^{(1-\phi)} \tag{2.67}$$

and for saturated, partially frozen soils

$$\lambda_{m,sat} = \lambda_w^{\theta_w} \lambda_i^{\theta_i} \lambda_s^{(1-\phi)} \tag{2.68}$$

where θ_w is the unfrozen volumetric water content, and θ_i is the volumetric ice content with the condition $\theta_w + \theta_i = \phi$. Thermal conductivity for wet material with water saturation S_w is interpolated according to Johansen (1975) as follows:

$$\lambda_m = \lambda_{m,\,dry} + (\lambda_{m,\,sat} - \lambda_{m,\,dry})Ke \tag{2.69}$$

where the interpolation factor Ke is referred to as Kersten's number (Kersten 1949) and is empirically approximated for coarse mineral soils as

$$Ke = 0.7 \log_{10} (S_w) + 1 \tag{2.70}$$

and for fine material

$$Ke = \log_{10} (S_w) + 1 \tag{2.71}$$

According to Nield and Bejan (2006) and Menberg et al. (2013a), the weighted geometrical model (Equation 2.67) provides good results as long as the thermal conductivities of water and solids are not too different from each other. Farouki (1981) compared Johansens's model with data from experiments with fine soils and found good correspondence for saturated and unsaturated conditions.

Balland and Arp (2005) extended Johansen's (1975) method in order to **model thermal conductivity over a wide range of conditions**, from loose to compact, organic to mineral, fine to coarse textured, frozen to unfrozen, and dry to wet. They retained Equation 2.69 for the thermal conductivity of wet material and Equations 2.67 and 2.68 for saturated unfrozen and partially frozen conditions. The thermal conductivity of solid material is calculated as

$$\lambda_s = \lambda_{organic}^{vf_{organic}} \lambda_{quartz}^{vf_{quartz}} \lambda_{mineral}^{vf_{mineral}} \tag{2.72}$$

where $vf_{organic}$, vf_{quartz}, and $vf_{mineral}$ are the volumetric fractions of organic material, quartz, and other minerals within the soil solids, respectively. Note that the thermal conductivity value for quartz is significantly higher than that of other typical minerals and organic matter. The **thermal conductivity for dry conditions** is determined as

$$\lambda_{dry} = \frac{(0.053\lambda_s - \lambda_a)\rho_b + \lambda_a\rho_p}{\rho_p - (1 - 0.053)\rho_b} \tag{2.73}$$

where ρ_p and ρ_b are the particle and the bulk densities. For Kersten's number Ke in Equation 2.69, they offer the new **model for unfrozen soils**:

$$Ke = S_{wi}^{0.5(1 + vf_{organic} - \alpha vf_{sand} - vf_{coarse})} \left[\left(\frac{1}{1 + \exp(-\beta S_{wi})} \right)^3 - \left(\frac{1 - S_{wi}}{2} \right)^3 \right]^{1 - vf_{organic}} \tag{2.74}$$

where $S_{wi} = S_w + S_i$; S_i is the ice saturation; vf_{sand} and vf_{course} are the volumetric fractions of the sand and the coarse grain size fraction, respectively; and α and β are adjustable parameters. By model calibration using extensive data sets, Balland and Arp (2005) determined the parameters to be $\alpha = 0.24 \pm 0.04$, and $\beta = 18.3 \pm 1.1$. Special importance has to be attached to the volumetric fractions $vf_{mineral}$ and vf_{quartz} within the soil solids. For frozen or partially frozen soil, Ke is

$$\text{Ke} = S_{wi}^{1+vf_{organic}} \tag{2.75}$$

The model is valid for the temperature range from approximately $-30°C$ to $30°C$. Data on the thermal properties and densities of the basic soil constituents used in the model are presented in Table 2.1.

Chen (2008) investigated the effect of porosity and saturation degree on thermal conductivity of quartz sands through laboratory tests. He found that thermal conductivity increases with the decrease in porosity and the increase in saturation degree. An empirical equation of thermal conductivity expressed as a function of porosity and saturation degree was developed:

$$\lambda_m(\phi, S_w) = \lambda_s^{(1-\phi)}\lambda_w^{\phi}\left[(1-b)S_w + b\right]^{c\phi} \tag{2.76}$$

For saturated conditions, it reduces to the weighted geometrical mean model. From fitting with experimental data, Chen (2008) found $b = 0.0022$ and $c = 0.78$. On the other hand, Johansen's (1975) method gave a fair correspondence with his data.

Further expressions are offered for **frozen soil conditions** (Johansen 1975; Farouki 1981). Values on thermal conductivity of water, sand packings, and aquifers are, besides Table 2.1, presented in Section 2.2.

Overall, **a variety of methods and techniques have been proposed for the effective thermal conductivity.** Which model should be used might

Table 2.1 Thermal properties and densities of basic soil constituents after Balland and Arp (2005) for the use of their modified Johansen's method

Soil component	Density (kg m^{-3})	Vol. heat capacity (J m^{-3} K^{-1})	Thermal conductivity (W m^{-1} K^{-1})
Quartz	2660	2.01×10^6	8.0
Other mineral	2650	2.01×10^6	2.5
Organic matter	1300	2.51×10^6	0.25
Water	1000	4.18×10^6	0.57
Ice	920	1.88×10^6	2.21
Air	0.00125	1.25×10^3	0.025

indeed be confusing for the practitioner. In our opinion, it is important to take note of the existing attempts. Nevertheless, we may consider the approach of Balland and Arp (2005) as a good starting point for practitioners. However, this does not prevent us from critically assessing its validity under prevailing physical conditions. Moreover, existing codes may already contain specific approaches for the effective thermal conductivity.

2.1.2.3 Dispersive and macrodispersive heat transport

Similar to solute transport processes, **mechanical dispersion effects** in porous media may also play a role in thermal processes provided that significant flow is present. In a homogeneous porous medium, mechanical dispersion effects are again due to the highly variable microscopic velocity field. The related fluxes are essentially advective heat fluxes. The average advective thermal flux can be expressed by average values:

$$\overline{qTC_w} = \overline{q}\,\overline{T}C_w + \overline{q'T'}C_w \tag{2.77}$$

where the overbar sign denotes the mean value within the control volume. The first term on the right-hand side corresponds to the advective thermal flux using mean values of specific water flux and temperature. Deviations from mean velocity and mean temperature within the control volume produce the **dispersive thermal heat flux** ($\overline{q'T'}C_w$), which is furthermore influenced by heat diffusion effects in the flowing water, as well as the matrix. The **thermal mechanical dispersion effect** can be explained as follows. Consider a cross section in a pore. A water particle in the middle of a pore is transported with a much higher velocity downstream than a water particle close to the solid wall. This causes a pronounced longitudinal spreading effect of the fluid and therefore of the temperature after some transport time. A transversal dispersion effect is mainly due to the lateral detours of the water particles around grains or solid blocks in the porous medium. Anyway, both effects are further affected by thermal diffusion effects, mainly lateral heat conduction, within the pores and the matrix.

The **specific thermal flux due to mechanical dispersion** in porous media is usually approximated in analogy to Fick's or Fourier's law by

$$j_{disp} = -\lambda_{disp}\nabla T \tag{2.78}$$

for saturated conditions (Bear 1972; Green et al. 1964), where λ_{disp} is analogous to the combined medium λ_m. For unsaturated conditions, it is correspondingly

$$j_{disp} = -\lambda_{disp}(S_w)\nabla T \tag{2.79}$$

The coefficient λ_{disp} is the **thermal conductivity tensor due to mechanical dispersion**. For isotropic porous media, the tensor is expressed by the **longitudinal and the transversal thermal conductivity coefficients due to dispersion**, $\lambda_{disp,\,L}$, and $\lambda_{disp,\,T}$ (W m^{-1} K^{-1}), where the longitudinal direction equals the flow direction and the transversal direction is normal to the flow direction. Often, the effects of thermal conduction and mechanical thermal dispersion are combined to the equivalent or **effective thermal conductivity tensor** λ_{eff}, which has the following components in longitudinal (L) and transversal (T) directions:

$$\lambda_{eff,L} = \lambda_m + \lambda_{disp,L}$$
$$\lambda_{eff,T} = \lambda_m + \lambda_{disp,T} \tag{2.80}$$

Note that the longitudinal dispersive flux component is formulated for the **mean direction of the groundwater flow**. Similarly, the transversal flux is expressed normal to the mean flow direction. If the x' axis of the coordinate system is chosen parallel to the flow direction in a **coordinate system with principal axes** x', y', z', the **effective thermal conductivity tensor** reads

$$\lambda_{eff} = \begin{bmatrix} \lambda_{eff,xx} & \lambda_{eff,xy} & \lambda_{eff,xz} \\ \lambda_{eff,yx} & \lambda_{eff,yy} & \lambda_{eff,yz} \\ \lambda_{eff,zx} & \lambda_{eff,zy} & \lambda_{eff,zz} \end{bmatrix} = \begin{bmatrix} \lambda_{eff,x'} & 0 & 0 \\ 0 & \lambda_{eff,y'} & 0 \\ 0 & 0 & \lambda_{eff,z'} \end{bmatrix} = \begin{bmatrix} \lambda_{eff,L} & 0 & 0 \\ 0 & \lambda_{eff,T,hor} & 0 \\ 0 & 0 & \lambda_{eff,T,vert} \end{bmatrix}$$

$$\tag{2.81}$$

Sometimes, the influence of thermal mechanical dispersion in homogeneous porous media is disregarded in theoretical studies (e.g., Bear 1972; Fujii et al. 2005; Woodbury and Smith 1985). This is motivated by the fact that the influence of thermal conductivity is often of similar magnitude to or larger than the thermal advection (Bear 1972; Woodbury and Smith 1985). Consequently, in thermal modeling, thermal dispersion is sometimes neglected (e.g., Domenico and Palciauskas 1973; Taniguchi et al. 1999; Reiter 2001; Ferguson et al. 2006).

The **effective thermal conductivity tensor** has been analyzed in a similar manner as the mechanical dispersion tensor for solute transport. Important parameters are the water velocity and the particle size of the porous medium (Green et al. 1964; Hsu and Cheng 1990; Levec and Carbonell 1985; Lu 2009; Metzger et al. 2004; Pedras and de Lemos 2008; Rau et al. 2012).

Metzger et al. (2004) investigated effective conductivity coefficients in laboratory experiments for packed beds (0.4 m long, 0.1 m wide) of glass beads (diameter 2 mm). They proposed the following correlation for the longitudinal and transversal effective thermal conductivity:

$$\lambda_{\text{eff,L}} = \lambda_m + b_L \lambda_w \text{Pe}_t^{m_L}$$
$$\lambda_{\text{eff,T}} = \lambda_m + b_T \lambda_w \text{Pe}_t^{m_T} \tag{2.82}$$

The dimensionless constant b_L is $b_L = 0.073$, the exponent m_L is $m_L = 1.59$, b_T is $b_T = 0.03–0.05$ (lower and upper limits), and $m_T = 1$. The dimensionless number Pe_t is the **thermal Peclet number**, which relates heat flux by advection to heat flux by conduction. It can be calculated as

$$\text{Pe}_t = \frac{C_w q d}{\lambda_m} \tag{2.83}$$

where d is a characteristic length, usually represented by the mean grain size. Equation 2.82 states that there is a **velocity dependence of the effective thermal conductivity coefficient**.

Rau et al. (2012) performed laboratory experiments on solute and heat transport in a homogeneous, well sorted quartz sand with a mean grain size of 2 mm. The chosen Darcy velocity range was 0.28 to 98 m day^{-1}. They found the following empirical formulae for the longitudinal and transversal effective thermal conductivity for their sand packing:

$$\lambda_{\text{eff,L}} = \lambda_m + \gamma_L \cdot C_m \left(\frac{C_w}{C_m} q \right)^2$$
$$\lambda_{\text{eff,T}} = \lambda_m + \gamma_T \cdot C_m \left(\frac{C_w}{C_m} q \right)^2 \tag{2.84}$$

with the factors $\gamma_L \cong 1.478$ s and $\gamma_T \cong 0.4$ s. The effective thermal conductivity coefficient therefore approximately depends on the square of the Darcy velocity. For the solute transport experiments, on the other hand, they found a linear dependence of the solute dispersion coefficient from the flow velocity. The longitudinal solute dispersivity was 3 mm. In their analysis, they showed that the key element is the thermal Peclet number (Equation 2.83), which ranged from 0.02 to 3 in their study. This range shows a distinct **transitional behavior**, where both thermal conduction and advection are of similar magnitude. For very small flow velocities, a constant thermal conductivity coefficient applies. On the other hand, a linear relationship, where thermal advection clearly dominates, can be expected for very high flow velocities with a thermal Peclet number of about 10 and higher. However, according to the authors, this is unrealistic for most practical applications. One has to keep in mind that the relationship 2.84 is valid only for the chosen homogeneous quartz sand packing. Nevertheless, it confirms the velocity dependence of the effective thermal conductivity tensor.

For geological formations, the dispersion phenomenon may be considerably affected and increased by **macrodispersion effects** due to the **high variability in hydraulic conductivity** (nonhomogeneous structures such as layers and lenses). Ferguson (2007) concluded from stochastic modeling based on geostatistical parameters that hydrodynamic macrodispersion is an important consideration in heat flow problems.

In order to **illustrate the macrodispersive heat flux** in a heuristic manner, we consider the simple extreme case of four horizontal layers as shown in Figure 2.8. The figure shows the vertical profile in the horizontal specific water flux $q(z)$ and the temperature profile $T(z)$ as a snapshot, assuming that flow of groundwater is horizontal and practically zero in two of the four layers with corresponding low temperature. Vertical heat flux is disregarded here. In this case, the total heat flux is easily calculated as $\overline{qT}C_w = 1$ unit, while the advective heat flux using mean hydraulic flux and mean temperature is $\overline{q}\,\overline{T}C_w = 0.5$ unit. Since the total flux $\overline{qT}$ can be expressed as $\overline{qT} = \overline{q}\,\overline{T} + \overline{q'T'}$, the macrodispersive flux can be calculated from the mean product of the deviations $q' = q - \overline{q}$, and $T' = T - \overline{T}$. The result is $\overline{q'T'}C_w = 0.5$ unit here. Therefore, the macrodispersive heat flux can be of the same order as the advective heat flux in extreme cases. This demonstrates the macrodispersion effect.

The **thermal macrodispersive flux** is again usually approximated by Equation 2.78. Similar to macrodispersion coefficients in solute transport, the **longitudinal and transversal thermal conductivity coefficients** are often

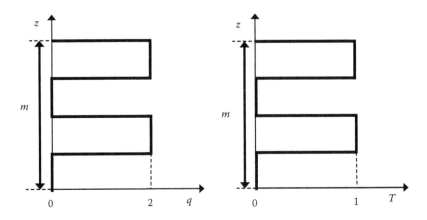

Figure 2.8 Example of a simple configuration with a layered aquifer and simple vertical specific flux and temperature profiles (schematic).

approximated by the Darcy velocity and by characteristic lengths in a linear manner:

$$\lambda_{disp,L} = \beta_L C_w q;$$
$$\lambda_{disp,T} = \beta_T C_w q \tag{2.85}$$

as suggested by Sauty et al. (1982) and adopted by de Marsily (1986). The parameters β_L (m) and β_T (m) are the **longitudinal and transversal thermal macrodispersivities**. The term $q = |q|$ is the absolute value of the Darcy flux. The evaluation of field experiments in 1978, for the Bonnaud aquifer site in France (single and double well system, transport distance approximately 13 m), revealed that the longitudinal dispersivities seemed to be comparable for solute and heat transport (Sauty et al. 1982). Therefore, it is often assumed in numerical modeling that $\beta_L = \alpha_L$ and $\beta_T = \alpha_T$, where α_L and α_T (m) are longitudinal and transversal macrodispersivities, respectively, for solute transport (Smith and Chapman 1983; Molson et al. 1992; Hopmans et al. 2002; Constantz 2008).

From **analogy with solute transport** in nonhomogeneous formations (e.g., Dagan 1989), it can be expected that the **thermal macrodispersivities** behave in a similar manner. The spatially variable hydraulic conductivity field $K(\mathbf{x})$ can be described by a covariance function which depends on the variance σ_Y^2 and the correlation length I_Y (integral scale) of the spatial variable $Y = \ln(K(\mathbf{x}))$. Scale dependence of field-scale macrodispersivity values of tracer transport in various aquifers was demonstrated by Gelhar et al. (1992). It states that macrodispersivity values start from local mechanical dispersivity and typically grow with increasing transport scale, thus exhibiting a pronounced **scale effect**.

Vandenbohede et al. (2009) found in the analysis of their field experiment that the longitudinal dispersivities for solute and heat transport are not comparable. They performed two push–pull tests, injecting chloride and cold water into an aquifer using a single well. The mean radial transport distances for the tracer were up to about 11 m. They found that thermal dispersivities do not seem to be scale dependent. In this context, one has to keep in mind that the mean transport distances for tracers and heat are different, which is due to the different mean velocities for the thermal and the solute front.

Molina-Giraldo et al. (2011) presented values of longitudinal thermal macrodispersivity from the literature versus field scale together with empirical and semiempirical relationships for solute transport (Neuman 1990; Xu and Eckstein 1995; Schulze-Makuch 2005) (Figure 2.9). The relationships have the following general form:

$$A_L = bL^{m_2} \tag{2.86}$$

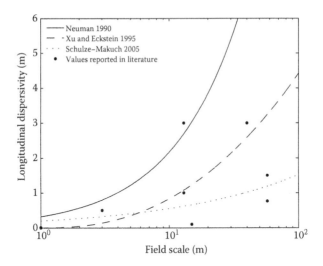

Figure 2.9 Longitudinal thermal macrodispersivity from the literature versus field scale together with empirical and semiempirical relationships for solute transport (Neuman 1990; Xu and Eckstein 1995; Schulze-Makuch 2005). (Modified after Molina-Giraldo, N. et al. *International Journal of Thermal Sciences* 50, 1223–1231, 2011.)

where b and m_2 are characteristic coefficients, and L is the length scale (advective transport distance). The data comprise evaluations from various unconsolidated aquifers worldwide. The longitudinal thermal dispersivity values clearly show scale effects. Measured values are best represented by the approach of Schulze-Makuch (2005). However, high variability for a given field scale is present, similar to the case of solute dispersivities shown by Gelhar et al. (1992). For a field scale of 10 m, for instance, the longitudinal dispersivity might be between 0.5 and 2 m. Nevertheless, most of the values are located within the ranges spanned by the empirical relationships. These differences reflect the specific geological conditions with respect to thermal dispersion effects.

Moyne et al. (2000) used stochastic concepts to express the development of the effective thermal conductivity in layered aquifers. **Hidalgo et al.** (2009) used a **stochastic approach** to describe heat transport in heterogeneous porous media, which is characterized by variance σ_Y^2 and correlation length I_Y. For steady-state conditions, longitudinal thermal dispersion is negligible, and the **transverse thermal dispersivity** is proportional to the variance of the log hydraulic conductivity Y and its correlation length I_Y, using a Gaussian covariance function as follows:

$$\beta_T = 0.02\sigma_Y^2 I_Y \tag{2.87}$$

As pointed out by Gelhar (1993), the correlation length I_Y of aquifers typically shows a distinct scale effect (typically 1/10 of the transport length scale).

Geiger and Emmanuel (2010) conducted numerical high-resolution finite element–finite volume simulations of heat transfer in two geologically realistic fractured porous domains. They calculated thermal breakthrough curves at various locations in the domains and analyzed them with a continuous time random walk (CTRW) model adapted for heat transfer. Their analysis shows that heat transport in the well-connected fracture network is Fourier-like, even though the thermal front is highly irregular. Consequently, it can be modeled by an advection–diffusion equation using macroscopic dispersivities. By contrast, heat transport in a poorly connected fracture pattern turned out to be highly non-Fourier-like. Hence, the classical advection–diffusion equation was not able to capture the main features, but they can be modeled successfully by CTRW. The authors conclude that the occurrence of non-Fourier behavior has important implications for a range of processes including geothermal reservoir engineering, radioactive waste storage, and enhanced oil recovery.

Chang and Yeh (2012) developed closed-form expressions using stochastic theory and the spectral perturbation techniques to describe the field-scale heat advection and the variability of the temperature profile in a partially saturated porous nonhomogeneous aquifer. Their results indicate that the macrodispersive heat flux depends on the spatial variability of the specific discharge, which, in turn, depends on the variation of hydraulic conductivity. The correlation length I_Y of log saturated hydraulic conductivity $Y(\mathbf{x})$ is important in enhancing the heat advection and the variability of the mean temperature field, and thus the macrodispersion effect. The longitudinal macrodispersivity value starts from a constant value and increases in an S-shaped manner over time or mean transport distance.

Nevertheless, the magnitude and development of the **effective thermal conductivity coefficient under field conditions are still a matter of debate** and have to be addressed in future research. In numerical models, however, **macrodispersion is usually taken into account with the help of constant (macro-)dispersivity values. Thermal dispersivity** values, which are only sparsely reported in the literature, are shown in Table 6.2.

A further physical effect, which is often modeled using a dispersion concept, is the **local temperature variation due to temporal variation of the flow field**, that is, nonpermanent flow direction and flow velocity. **Numerical simulations with solute transport in transient flow fields have** shown that considerable quasi-dispersive effects may occur. Kinzelbach and Ackerer (1986) showed that transversal dispersivity in contaminant transport is increased, and longitudinal dispersivity decreased in transient flow fields compared to steady-state ones. Dentz and Carrera (2005) showed for heterogeneous hydraulic conductivity fields that stochastic longitudinal and transversal fluctuations of the hydraulic gradient can contribute to the effective dispersion coefficient even more strongly than macrodispersion does. It can be expected that **effects of transient flow fields are also relevant in heat transport**. Particularly long thermal plumes, which are modeled using mean velocity field, may thus exhibit considerable additional transversal dispersion effects.

2.1.2.4 Heat transport equation

Based on the expressions for storage of heat, advective heat flux, and heat conduction, including thermal macrodispersion effects and assuming that the mean temperatures of the water and the solid phase are the same within the control volume, the **energy balance for a unit volume of saturated or unsaturated porous medium** (Figure 2.4) can be formulated as

$$C_m \frac{\partial T}{\partial t} = \nabla \cdot \left[(\lambda_m + \lambda_{disp}) \nabla T \right] - C_w \nabla \cdot (\mathbf{q} T) + P_t \tag{2.88}$$

where P_t is a thermal production term (heat production per unit volume and unit time) (W m^{-3}). The equation states that the rate of change of energy content equals the energy inflow minus the outflow over a unit control volume increased by the energy production in that volume. The assumption of equal mean temperature in water and solids of porous media is not exactly true at the microscopic level (Moyne et al. 2000).

If **freezing and thawing** are considered, the heat transport equation can be reformulated as follows (Williams and Smith 1989; Hansson et al. 2004):

$$C_m \frac{\partial T}{\partial t} - L_f \rho_i \frac{\partial \theta_i}{\partial t} = \nabla \cdot \left[\lambda(\theta_l, \theta_i) \nabla T \right] - C_w \nabla \cdot (\mathbf{q}_w T) + P_t \tag{2.89}$$

where L_f (J kg^{-1}) is the latent heat of melting/freezing. The second term in Equation 2.89 represents the energy needed to melt the ice mass $\rho_i \theta_i$ per unit time and vice versa in the case of freezing. Note that this amount of energy is quite large (3.34×10^5 J kg^{-1}). In this case, the thermal conductivity also depends on the volumetric ice content. Here, it is assumed that the heat transport by air flow can be neglected. Still, a relationship is needed to fully define the systems. It is obtained by the **relationship between the liquid water pressure and temperature**, $p_w(T)$, for the absolute temperature $T < T_0$ (K), where T_0 is the freezing point temperature, in approximate form:

$$\frac{p_w}{\rho_w g} \simeq \frac{L_f (T - T_0)}{g T_0} \tag{2.90}$$

It is an application of the **principles of Clausius–Clapeyron on phase change parameters**. The **corresponding liquid water content** for freezing conditions is approximated using the relation between capillary pressure (i.e., water pressure p_w here with $p_a = 0$) and water saturation $S_w(p_w)$ of volumetric water content. For example, after Brooks and Corey (1966), the relation $\theta_w(T)$ is

$$\theta_w(T) = \theta_{w,r} + (\phi - \theta_{w,r}) \left(\frac{h_b T_0 g}{(T_0 - T)} \right)^{\lambda_{BC}}$$

$$\text{for} \quad \frac{(T_0 - T)L_f}{T_0 g} > h_b$$

(2.91)

Although the liquid water saturation is extremely small for frozen conditions, it is still present.

Dividing the energy balance Equation 2.88 (without freezing/thawing) by the volumetric thermal capacity C_m and combining the effects of heat conduction and thermal macrodispersion in a thermal dispersion tensor $\mathbf{D}_t$ yields the **heat transport equation**

$$\frac{\partial T}{\partial t} = \nabla \cdot [\mathbf{D}_t \nabla T] - \frac{C_w}{C_m} \nabla \cdot (\mathbf{q}T) + \frac{P_t}{C_m}$$

(2.92)

Adopting a linear dependence of the macrodispersion coefficient on Darcy velocity, the **thermal diffusivity tensor** $\mathbf{D}_t$ in a hydraulically isotropic medium has the principal **longitudinal and transversal components**:

$$D_{t,L} = D_{t,\text{diff}} + D_{t,\text{disp},L} = \frac{\lambda_m}{C_m} + \frac{\beta_L C_w |\mathbf{q}|}{C_m};$$

$$D_{t,T} = D_{t,\text{diff}} + D_{t,\text{disp},T} = \frac{\lambda_m}{C_m} + \frac{\beta_T C_w |\mathbf{q}|}{C_m};$$

(2.93)

The **thermal diffusion coefficient** $D_{t,\text{diff}} = \lambda_m/C_m$ (m² s⁻¹) of a porous medium representing solely the effect of heat conduction is typically of the order of 10^{-7} to 10^{-6} m² s⁻¹ for granular aquifers. This means that it is up to three orders of magnitude larger than the molecular diffusion coefficient. Therefore, thermal mechanical dispersion for homogeneous porous media is usually not dominant in heat transport problems. It may be of similar order of magnitude as thermal diffusion. Nevertheless, the effect of macrodispersion in nonhomogeneous aquifers is still present, as stated above, depending on the prevailing macrodispersivity values. For typical situations, the macrodispersive flux dominates over the thermal diffusive flux due to heat conduction. Consequently, we keep the tensorial form of the heat transport equation.

We may compare the heat transport equation with the transport equation for a dissolved species with linear sorption and first order decay:

$$R_c \frac{\partial c}{\partial t} = \nabla \cdot \left[\mathbf{D}_h \nabla c \right] - \nabla \cdot [c\mathbf{u}] - \lambda_c R_c c \tag{2.94}$$

where R_c is the retardation factor (accounting for linear sorption effects), c (kg m^{-3}) is the solute concentration, $\mathbf{D}_h$ is the hydrodynamic dispersion coefficient (tensor) (m^2 s^{-1}), $\mathbf{u}$ (m s^{-1}) is the mean flow velocity vector ($\mathbf{u} = q/\phi$), and λ_c (s^{-1}) is the decay coefficient. From the comparison, we can see that the variable T corresponds to c, the thermal diffusion tensor to the hydrodynamic dispersion tensor. The mean water velocity $\mathbf{u}$ corresponds to the **thermal velocity** $\mathbf{u}_t$ (m s^{-1}):

$$\mathbf{u}_t = \mathbf{q} \, C_w/C_m \tag{2.95}$$

This relation can be obtained by the advective thermal flux condition in porous media:

$$\mathbf{q}C_w \, (T - T_0) \, \mathbf{u}_t C_m \, (T - T_0) \tag{2.96}$$

As a consequence, when advective heat transport is dominant over heat conduction, a thermal front in groundwater propagates with the thermal velocity:

$$u_t = \frac{C_w \phi}{C_m} u \tag{2.97}$$

which is typically 2 to 3 times smaller than that for solute transport in granular aquifers. Therefore, a thermal front is retarded with respect to an ideal tracer front, with a **thermal retardation factor** R_{t_ret} [-] of

$$R_{t\_ret} = \frac{C_m}{\phi C_w} \tag{2.98}$$

Shook (2001) found that the **ratio of water to temperature velocity is constant, even for heterogeneous porous media.** Therefore, thermal breakthrough in heterogeneous media can be predicted from tracer tests. However, in the presence of strong permeability correlations, like in the case of layering, some deviations may occur.

Lo Russo and Taddia (2010) showed the prevalence of heat advective transport with respect to thermal dispersion for their field site in Torino, Italy. They investigated advective heat transport induced by the injection of an open-loop system. Thermal stratification was explained by prevailing horizontal advection of the flowing groundwater.

For **unsaturated conditions**, the porosity ϕ in Equations 2.97 and 2.98 has to be replaced by the water content $\theta_w = \phi S_w$. Note that according to Equation 2.46, the volumetric thermal capacity C_m depends on the water saturation S_w as well, which can be obtained from the solution of the flow problem. If uniform vertical flow conditions are assumed, the specific discharge q is constant and the vertical flow gradient is $I_{vert} = 1$ for homogeneous soils. We can therefore determine S_w using Equation 2.14 by setting $q = K_w(S_w)$ in this case. Unsaturated conditions can have quite some impact on the thermal front velocity.

For the modeler, it is important to know whether thermal macrodispersion dominates over heat conduction, or whether heat conduction dominates over heat advection. In the first case, the **ratio between dispersive and advective heat fluxes** can be expressed by the **dimensionless number**

$$\frac{\text{thermal dispersive flux}}{\text{flux by heat conduction}} = \frac{\beta_L C_w q}{\lambda_m} \tag{2.99}$$

Depending mainly on the longitudinal macrodispersivity and the Darcy flux values, the ratio indicates dominance of the respective term. The second case is described by the **thermal Peclet number** (Equation 2.83). For $Pe_t \gg 1$, heat advection effects dominate over heat conduction. However, for **very small flow velocities** with $Pe_t \ll 1$, the heat transport Equation 2.92 reduces to the **heat conduction equation**:

$$C_m \frac{\partial T}{\partial t} = \nabla \cdot [\lambda_m \nabla T] + P_t \tag{2.100}$$

It can be written in the form of a diffusion equation (Carslaw and Jaeger 1959):

$$\frac{\partial T}{\partial t} = \nabla \cdot [D_t \nabla T] + P_t \tag{2.101}$$

where D_t is the thermal diffusion coefficient (scalar) with $D_t = \lambda_m / C_m$.

Characteristic quantities of thermal propagation, therefore, depend on the thermal Peclet number Pe_t. For very small Pe_t, the coefficients λ_m and C_m are needed. Characteristic parameters for larger Pe_t numbers are the **specific flux field** $q(\mathbf{x}, t)$ (Darcy flux), the **thermal capacity ratio** C_w/C_m, and **the thermal diffusivity tensor** $\mathbf{D}_t$. The latter is typically dominated by macrodispersive effects characterized by the **longitudinal and transversal macrodispersivities** β_L and β_T. The ratio C_w/C_m is about 1.8 for granular aquifers. The **thermal conductivity** of porous materials λ_m is still needed to express boundary fluxes.

2.1.2.5 Initial and boundary conditions

The **initial condition for a heat transport problem** consists of a specified temperature distribution $T(\mathbf{x}, t = 0)$ for the whole solution domain.

Boundary conditions for heat transport problems are essentially specified values of the temperature T at a boundary section B_1, a specified conductive or convective–conductive heat flux through a boundary section B_2 or B_3. A further boundary consists of the outflow boundary B_4. All these sections are part of the boundary.

Specified temperature T_{B1} at a boundary section B_1 (first type or Dirichlet boundary condition; Figure 2.10a) is expressed by

$$T\left(\mathbf{x}_{B_1}, t\right) = T_{B_1}\left(\mathbf{x}_{B_1}, t\right); \quad \mathbf{x}_{B_1} \in B_1 \tag{2.102}$$

An example for B_1 is the infiltration zone from a surface water body at water temperature T_1 into the saturated or the unsaturated zone, or a borehole heat exchanger (BHE) with specified temperature. Another example is a soil surface with specified temperature. In this context, we have to be aware that the temperature at the soil surface, that is, just within the soil, is the result of a complex energy balance, taking into account input from shortwave and longwave radiation and outflowing longwave radiation (see Section 1.2.1), as well as fluxes into the ground caused by BHEs. Moreover, evaporation and transpiration effects occur as well as convective interactions with air and vapor flow close to the soil surface. Taking the resulting **soil surface temperature**, based on measurements, leads to a simplified formulation of the complex situation. It can be estimated using the relation with air temperature, as indicated in Section 1.2.1.

Specified conductive heat flux $j_{2,n}$ (normal component) through a boundary section B_2 requires (second type or Neumann boundary condition; Figure 2.10b)

$$j_n\left(\mathbf{x}_{B_2}, t\right) = -\lambda_n \left(\frac{\partial T}{\partial n}\right)_{B_2} = j_{2,n}\left(\mathbf{x}_{B_2}, t\right); \quad \mathbf{x}_{B_2} \in B_2 \tag{2.103}$$

The flux j_n is the thermal flux in the normal direction to the boundary surface, which is oriented from the solution domain to the outside. This means that for positive temperature gradients, it represents a thermal flux

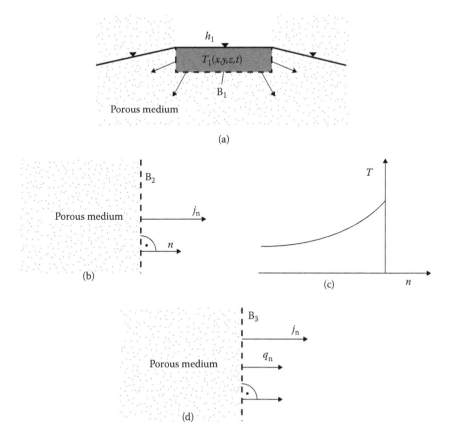

Figure 2.10 Thermal boundary conditions (schematic). (a) Surface water body with prescribed temperature, B_1; (b) specified conductive flux boundary, B_2; (c) temperature profile for solid material in contact with an outside fluid; (d) specified advective-conductive heat flux boundary, B_3.

out of the solution domain. An example for B_2 is a zone of specified conductive heat flux through the aquifer bottom, for example, by geothermal heat flux. A **thermally insulating boundary** is obtained by setting the condition $j_{2,n} = 0$.

For **solid materials** (e.g., solid wall) **in contact with an outside fluid** (e.g., air, or water), a mixed type (or third type or Cauchy boundary condition) is often applied, similar to Figure 2.10b and c (e.g., Gröber et al. 1955):

$$j_n\left(\mathbf{x}_{B_2},t\right)=-\lambda_n\cdot\left(\frac{\partial T}{\partial n}\right)_{B_2} = \alpha_t\cdot\left(T\left(\mathbf{x}_{B_2}\right)-T_{B_2}\right); \quad \mathbf{x}_{B_2} \in B_2 \qquad (2.104)$$

where T_{B2} is the temperature of the fluid outside of the thin thermal boundary layer close to the boundary, and α_t is the heat transfer coefficient (W K^{-1} m^{-2}). The latter coefficient depends on the flow conditions and state of the fluid.

Specified **advective–conductive heat flux** $j_{3,n}$ (normal component) through a boundary section B_3 requires (third type or Cauchy boundary condition; Figure 2.10d)

$$j_n\left(\mathbf{x}_{B_3}, t\right) = q_n(T - T_0) - \lambda_n\left(\frac{\partial T}{\partial n}\right)_{B_3} = j_{3,n}\left(\mathbf{x}_{B_3}, t\right); \quad \mathbf{x}_{B_3} \in B_3 \qquad (2.105)$$

An example for B_3 is an inflow face of an aquifer with specified advective–conductive heat flux, or the borehole surface of a heat exchanger system where the heat flux is given. The role of the advective heat flux of infiltrating water is discussed in Kollet et al. (2009).

A further type of boundary condition concerns the **heat flux through the groundwater outflow face** of an aquifer (aquifer section B_4). The usual assumption of the heat insulation condition $(\partial T/\partial n) = 0$, where n is the normal direction to the outflow face, is often considered as unsatisfactory. An alternative is the establishment of the so-called **transmission boundary condition**, as sometimes used in solute transport models. The condition requires that the temperature gradient across the outflow boundary remains constant. This condition can be fulfilled in an approximate manner in numerical models by setting the dispersivity values to zero in the boundary cells.

In this context, we may pose the question about the **adequate thermal boundary conditions for technical systems** for the thermal use of aquifers. In the case of **open systems** with a defined inflow rate and temperature, the total inflowing borehole heat flux J_{bt} (W) (heat load) is specified. The specific heat flux can be related to a unit borehole length or borehole area.

For **closed systems** with BHEs, the steady-state total **borehole heat load** J_{bt} (W or J s^{-1}) is usually calculated using the mean temperature T_f (K) of the circulating fluid, the mean borehole surface temperature T_b (K), the length L (m) of the exchanger system, and a thermal borehole resistance R_t (K W^{-1} m^{-1}) as follows (e.g., Lamarche et al. 2010):

$$J_{bt} = \frac{L(T_f - T_b)}{R_t} \qquad (2.106)$$

Various authors have proposed analytical and empirical approaches for the determination of the **thermal borehole resistance** (Lamarche et al. 2010; Wagner et al. 2013). The thermal resistance R_t depends on the specific geometrical configuration of the heat exchanger system (like vertical single U-tube borehole embedded in grouting material; Figure 1.8), the thermal conductivity of soil, pipe, and borehole grouting material. Advective ground-water flow is usually disregarded. However, a recent study by Wagner et al. (2013) shows how groundwater-influenced thermal response tests (TRTs) in grouted BHEs with Darcy velocities > 0.1 m day^{-1} can be analyzed. The latter is comprehensively discussed in Chapter 6. Based on Equation 2.106 and depending on the mode of operation of the heat pump system, a **speci-fied borehole temperature** or a **specified heat flux** at the borehole may be appropriate boundary conditions for the heat transport model. As pointed out by Wang et al. (2012), the **thermal performance** of heat exchanger sys-tems in aquifers **with groundwater flow** may be strongly increased. They concluded from their case studies that the enhanced effect of the groundwa-ter flow depends greatly on the amount, thickness, and depth of aquifers. The effect will mainly affect the mean borehole surface temperature T_b.

2.1.2.6 Concepts for BHEs

In order to implement the specific **design of BHEs** (single- or double-U-tube, coaxial BHE) into numerical models, various mathematical methods have been proposed. Since a detailed modeling of the complex three-dimensional BHE-aquifer systems including all processes in the inflowing and outflow-ing tubes and inside the borehole is cumbersome and time-consuming in practical applications, various concepts have been proposed in the past. Nevertheless, detailed numerical analyses have been performed for com-parison and test purposes.

Consider a **single U-tube configuration** according to Figure 2.11. Both legs of the U-tube exhibit a different mean temperature within the tube, $T_{f1}(z, t)$ and $T_{f2}(z, t)$. The temperature at the borehole surface is $T_b(z, t)$

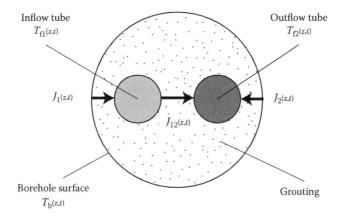

Figure 2.11 Single tube configuration with inflow and outflow tube. Schematic cross sec-tion at level z along the BHE.

neglecting axial temperature variation. The prevailing heat fluxes per unit length increment Δz are the heat fluxes from the borehole surface to each of the tubes and the heat fluxes between the tubes.

A widely used concept consists of relating the **heat flux from a cylindrical borehole wall to one single pipe**, which is embedded in grout material, to the **difference between the surface temperature of the borehole and the temperature of the fluid** circulating in the tube, together with a unit length thermal resistance R_t (see Section 2.1.2.5 and Equation 2.106). In the case of pipe 1, the heat flux is

$$J_1(z,t) = \frac{\left(T_b(z,t) - T_{f1}(z,t)\right)\Delta z}{R_{t1}} \tag{2.107}$$

where R_{t1} (K m W^{-1}) is the thermal resistance of pipe 1 including the grouting, and Δz (m) is a length increment along the borehole. The inverse of R_{t1} is related to a heat transfer coefficient. The other fluxes J_2 and J_{12} are formulated in a similar manner, with thermal resistance R_{t2} and R_{t12}. Equation 2.107 represents **a two-dimensional concept** in the plane normal to the axis of the BHE, valid for a particular location z along a vertical BHE at time t.

For such a single U-tube configuration, Eskilson and Claesson (1988) formulated the **heat balance in both tubes** as follows:

$$\rho_f c_f Q_f \frac{\partial T_{f1}}{\partial z} = \frac{J_1}{\Delta z} - \frac{J_{12}}{\Delta z} = \frac{T_b - T_{f1}}{R_{t1}} - \frac{T_{f1} - T_{f2}}{R_{t12}}$$

$$-\rho_f c_f Q_f \frac{\partial T_{f12}}{\partial z} = \frac{J_2}{\Delta z} + \frac{J_{12}}{\Delta z} = \frac{T_b - T_{f2}}{R_{t2}} + \frac{T_{f1} - T_{f2}}{R_{t12}} \tag{2.108}$$

where Q_f (m^3 s^{-1}) is the pumping rate of the circulating fluid (heat carrier fluid). Heat capacity effects of all materials within the borehole and heat conduction in the axial direction are neglected. Equation 2.108 is valid along the vertical borehole and represents a coupled differential equation system. The **initial condition** is given by constant soil surface temperature and constant geothermal gradient. The **boundary condition** of the system is given by constant surface temperature and by specifying the inflow temperature $T_{f1}(z = 0, t)$. The constant soil surface temperature and the consideration of further boreholes are considered by using the superposition principle. The two pipes are connected at the **bottom of the borehole**, thus requiring equal temperature there. The term on the left-hand side in both equations denotes the rate of change of heat within the pipe per unit length of pipe due to heat advection within the pipe. The terms on the right-hand side are the heat fluxes from the borehole wall to the pipes and the heat flux between the pipes. All these heat fluxes are taken as quasi-steady fluxes, thus disregarding heat capacity effects within the borehole. Therefore, the **timescale** has to be larger (Eskilson and Claesson 1988) than the timescale for thermal diffusion in the borehole and larger than the time needed for exchanging the fluid mass in the pipes:

$$\Delta t \geq \frac{5r_b^2}{D_t} + \frac{\pi r_p^2 2H}{Q_f} \tag{2.109}$$

where r_b (m) is the radius of the borehole, r_p (m) is the radius of the pipes, D_t is the thermal diffusion coefficient, and H is the length of the BHE. The **total heat flux along the borehole** can be expressed by

$$J_b(z,t) = \left(\frac{T_b - T_{f1}}{R_{t1}} + \frac{T_b - T_{f2}}{R_{t2}} \right) \Delta z \tag{2.110}$$

This heat flux along the borehole has to be coupled with the global heat transport outside the borehole.

Analytical solutions to equation system 2.108 are presented by Eskilson and Claesson (1988). The solution for $T_{f1}(z, t)$ and $T_{f2}(z, t)$ are related to the inflow temperature $T_{f1}(z = 0, t)$ and outflow temperature $T_{f2}(z = 0, t)$ of the fluids in the two pipes and the temperature profile $T_b(z, t)$ along the borehole. The temperature $T_{f2}(z = 0, t)$ yields the outflow temperature from the BHE.

Zeng et al. (2003) developed analytical solutions for the temperature profiles in the legs of single and double-U-tube BHE. They assumed that the borehole wall temperature T_b is invariant along the borehole depth but may change over time. Solutions (depending on T_1, T_2, and T_b) were developed for various combinations of the double U-tube for parallel and serial configurations.

Yang et al. (2009) combined the outside and inside regions of single U-tube configuration in an iterative manner. For the outside region, they used a cylindrical source model (Section 3.1.4).

For the **single, double U-tube, and the coaxial tube configurations**, related mathematical models were formulated by Diersch et al. (2011). They included transient heat storage, as well as thermal dispersion within the pipes. Still the thermal resistance concept is adopted. The grout material zone was subdivided into two (single U-tube) and four (double U-tube) subzones with corresponding grout temperatures. For quasi steady-state heat flux within the borehole, they provided **analytical solutions** analogous to Eskilson and Claesson (1988). According to Diersch et al. (2011), the analytical solutions strategy in the overall solution is highly efficient, precise, and robust. However, it is restricted to long-term processes, with timescales of the order of hours. They usually consider this limitation of the analytical method to be irrelevant for real BHE applications, where the thermal process scales are measured in days and years. If finer temporal resolutions are required, a **numerical treatment of the transient system**, which is embedded in their finite element formulation for the aquifer, is feasible. Bauer et al. (2011) developed thermal resistance and capacity models for different types of BHEs. By considering the thermal capacity of the grouting material, a higher accuracy can be reached in transient simulations. This can be important, for example, in the case of TRTs. Bauer et al. (2011) checked their model against simulations using fully discretized finite element models. Zarrella et al. (2011) also extended the model for double U-tube configurations by considering thermal capacity of the grouting material in order to account for short-term analyses.

The unit length thermal resistance R_t depends on the geometrical configuration of the BHE, the thermal conductivity of the grout and pipe wall material, and heat conduction as well as heat convection effects within the pipe. It can be determined experimentally, empirically, or numerically or by analytical approximations. A review of thermal resistance formulations for the example of single U-tube configurations is given by Lamarche et al. (2010). Analytical expressions for the thermal resistance of single and double U-tube and coaxial configurations can be found in Diersch et al. (2011). Hellström (1991) and Claesson and Hellström (2011) developed the so-called multipole method to evaluate the thermal resistances between the heat carrier fluid in the pipes of the borehole and the immediate vicinity of the surrounding ground. Sagia et al. (2012) concluded from their numerical analysis that the borehole thermal resistance decreases as the spacing between GHE pipes increases, and that a rise in the thermal conductivity of the grout material leads to a decrease in the borehole resistance. Furthermore, a decrease in the pipe's diameter enables a decrease in the thermal resistance between the heat carrier fluid and the ground, and a small value of borehole thermal resistance is desirable in order to achieve a high performance of BHE systems. Based on their analytical model and Hellström's (1991) multipole solution for the thermal resistance, Zeng et al. (2003) expressed the effective borehole resistance. Their calculations show that the double U-tube boreholes are superior to those with the single U-tube with respect to the overall thermal resistance and that double U-tubes in parallel configuration show better performance than those in series. Jun et al. (2009) compared several thermal resistance models with data from a field study in Shanghai (China). They found that in their case, short-term thermal resistance is about 76% of the long-term resistance. Line-source and cylindrical source theory were successful as long as the thermal processes were conduction dominated. Based on a two-dimensional numerical analysis, Sharqawy et al. (2009) developed a correlation to express the effective borehole resistance, which deviates from current semianalytical models.

Marcotte and Pasquier (2008) showed in their numerical analysis that the **thermal resistance in borehole thermal conductivity tests** is overestimated when using the usually applied average temperature of the fluid entering and leaving the ground. They instead proposed a new estimator they termed "*p*-linear" using temperature variations to a power of $p \to -1$. The proposed *p*-linear average is

$$\left|\Delta T_p\right| = \frac{p\left(\left|\Delta T_{in}\right|^{p+1} - \left|\Delta T_{out}\right|^{p+1}\right)}{(1+p)\left(\left|\Delta T_{in}\right|^{p} - \left|\Delta T_{out}\right|^{p}\right)} \tag{2.111}$$

Sutton et al. (2003) developed a new ground resistance model for vertical BHEs with groundwater flow.

2.1.2.7 Coupling thermal transport with hydraulic models

In the case of **strong variability of water density and viscosity**, due to temperature fluctuations, the **water flow and the heat transport equations have to be solved simultaneously** in a coupled manner, because temperature changes, in general, affect water flow processes via temperature dependence on water density and viscosity. The related phenomenon of **density-driven flow** is **thermal convection** in porous media (e.g., Nield and Bejan 2006).

Considering the **flow Equation 2.24** for saturated aquifers:

$$S_s \frac{\partial h_w}{\partial t} + \phi b_T \frac{\partial T}{\partial t} = \nabla \cdot \left[K_w(T)(\nabla h_w + b_T \nabla z \cdot (T - T_0)) \right] + \frac{\rho_w(T)}{\rho_{w,0}} w \tag{2.112}$$

and the **heat transport Equation 2.92**:

$$\frac{\partial T}{\partial t} = \nabla \cdot \left[D_t \nabla T \right] - \frac{C_w}{C_m} \nabla \cdot (qT) + \frac{P_t}{C_m} \tag{2.113}$$

It becomes evident that the head $h_w(x)$ and the flow field $q(x)$ are temperature dependent and exhibit **possible density effects**, which in turn are used to evaluate heat transport. Nonlinearities exist in the temperature-dependent hydraulic conductivity $K_w(x, T)$ and the water density $\rho_w(x, T)$.

For **small density differences**, the flow and the heat transport equations can be solved in an **uncoupled, sequential** manner.

Another type of coupling occurs at the soil surface. Parlange et al. (1998) demonstrated the **importance of advective water vapor transport to the mass and energy balance of diurnally heated soil surfaces of field soils**. This flux arises from the expansion and contraction of the soil air due to heating and

cooling over the day. Their analysis requires a coupled unsaturated water flow
and a heat transport model that takes water vapor flux into account.

2.1.2.8 Two-dimensional heat transport models

The heat transport equation for **shallow regional aquifers** can be obtained
by vertically integrating the three-dimensional equation according to the
hydraulic case described in Section 2.1.1.2. The corresponding control volume extends, again, from the aquifer bottom to the water table according to Figure 2.6. Besides the horizontal thermal fluxes, the vertical fluxes
through the bottom face and the top face of the control volume also have
to be taken into account. These fluxes consist of the heat conduction flux
through the bottom face and both the heat conduction and heat advection fluxes through the top face. However, because the fluxes through the
top face will be highly transient over the year and difficult to express in
detail, they might be represented in a simpler manner by considering yearly
average conditions. Under these assumptions, **steady-state heat transport in
shallow regional aquifers** (Figure 2.12) can be approximated by

$$\nabla \cdot (\mathbf{D}_t \nabla T) - \frac{C_w}{C_m} \nabla \cdot (\mathbf{q} T) + \frac{P_t}{mC_m} + \frac{j_{\text{vert,bot}}}{mC_m} + \frac{j_{\text{vert,top}}}{mC_m} = 0 \qquad (2.114)$$

The parameter m is the aquifer thickness. The vertical thermal flux
through the bottom face might be directly specified, possibly by the geothermal heat flux. The specific **vertical thermal flux through the top face**
of the control volume can be approximated by linear expressions for the
related conductive and advective fluxes:

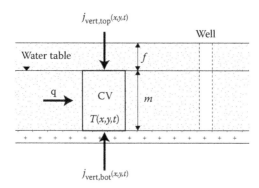

Figure 2.12 Schematic cross section through a shallow, extended unconfined aquifer,
with control volume (CV) extending from aquifer bottom to the water
table; thermal conditions.

$$j_{\text{vert,top}} \simeq \frac{\lambda_{\text{vert}}(T_{\text{surface}} - T)}{(f + m/2)} + NC_w(T_{\text{surface}} - T_0) \tag{2.115}$$

where λ_{vert} is an effective thermal conduction, expressing the thermal conductivity structure between soil surface and mean level of the saturated zone. The total thermal transfer distance therefore consists of the depth f to groundwater, and half the aquifer thickness, that is, $m/2$. The advective term depends on the mean (natural) recharge rate N (accretion rate, volume per unit area, and unit time).

Assuming, **for example**, an effective vertical thermal conductivity of $\lambda_{\text{vert}} = 2$ W m^{-1} K^{-1}, a depth to groundwater f of 5 m, an aquifer thickness m of 10 m, and a ΔT of 3 K, the related vertical conductive heat flux is $j_{\text{cond}} = 0.6$ W m^{-2}. The vertical advective heat flux j_{adv}, on the other hand, is for a mean recharge rate of 1 mm day^{-1} and the same ΔT, $j_{\text{adv}} = 0.15$ W m^{-2}. Therefore, vertical flux by heat conduction is dominant in this example.

2.1.3 Integral water and energy balance equations for aquifers

Integral balance equations for water and heat represent interesting tools for assessing and identifying the most important contributors in the context of the thermal use of shallow aquifers. However, in general, they do not replace the formulation of mathematical models, as described in Sections 2.1.1 and 2.1.2. Consider an aquifer domain D with saturated and possibly unsaturated zones. An integral water balance relates the rate of change of water volume within the domain D to all water fluxes in and out of it:

$$\frac{dV_w}{dt} = \sum_i^{M_{w\_in}} Q_{i\_in}(t) - \sum_i^{M_{w\_out}} Q_{i\_out}(t) \tag{2.116}$$

where $V_w(t)$ (m^3) is the water volume within the aquifer domain D, $Q_{i\_in}(t)$ (m^3 s^{-1}) are the inflow rates, $Q_{i\_out}(t)$ are the outflow rates (≤ 0), and M_{w_in} and M_{w_out} are the number of the inflowing and outflowing water flow components, respectively. A schematic example for an unconfined aquifer is shown in Figure 2.13. **Hydraulic inflow and outflow rates** Q_i can be listed as follows:

- Inflow $Q_{\text{sw\_in}}(t)$ from surface water bodies, like rivers and lakes
- Outflow $Q_{\text{sw\_out}}(t)$ from groundwater into surface water bodies
- Inflow $Q_{\text{replenish}}(t)$ from distributed natural replenishment at soil surface contributing to aquifer recharge
- Inflow $Q_{\text{lat\_in}}(t)$ through inflow into the aquifer from upstream regions, and lateral boundaries, including the aquifer bottom into the saturated or unsaturated zones of D

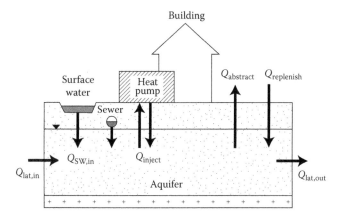

Figure 2.13 Hydraulic inflow and outflow rates for a schematic unconfined aquifer.

- Outflow $Q_{\text{lat\_out}}(t)$ from the aquifer into downstream regions, and from lateral boundaries including the aquifer bottom
- Water injection (inflow) $Q_{\text{inject}}(t)$, for example, by recharge wells, or infiltration by leaky sewers, or infiltration of roof water, or infiltration for thermal use
- Water abstraction (outflow) $Q_{\text{abstract}}(t)$, for example, by wells for drinking or process water use, or for thermal use

Note that **inflow is positive** and **outflow is negative**.

The **development of the integral water volume** in the aquifer over time can be formulated as follows:

$$V_{\text{w}}(t) = V_{\text{w}0}(t_0) + \sum_{i=1}^{M_{\text{w\_in}}} \int_{t_0}^{t} Q_{i\_\text{in}}(t)\,\mathrm{d}t + \sum_{i=1}^{M_{\text{w\_out}}} \int_{t_0}^{t} Q_{i\_\text{out}}(t)\,\mathrm{d}t \qquad (2.117)$$

where $V_{\text{w}0}$ is the initial water volume in the aquifer. In the **long run,** neglecting water volume changes for large values of time t, the water balance is simply

$$\sum_{i=1}^{M_{\text{w\_in}}} \overline{Q_{i\_\text{in}}} + \sum_{i=1}^{M_{\text{w\_out}}} \overline{Q_{i\_\text{out}}} = 0 \qquad (2.118)$$

where $\overline{Q_i}$ are time-averaged rates, with

$$\overline{Q_i} = \frac{1}{t - t_0} \int_{t_0}^{t} Q_i(t)\,\mathrm{d}t \qquad (2.119)$$

The rates $Q_i(t)$ have to be determined or estimated based on measurements or by modeling. The replenishment rate, for example, is calculated based on precipitation measurements minus modeled evapotranspiration (e.g., after Allen et al. 2006) and minus surface runoff. The flow from or into surface water bodies, with all the interactions between surface water and aquifer, is, in general, difficult to assess and usually requires two- or three-dimensional modeling using a calibrated groundwater flow model. Since we are primarily interested in average flow rates $\overline{Q}_i$, a steady-state model using time-averaged boundary conditions may be sufficient. The same holds true also for the lateral outflow rates from an aquifer, while inflow rates may be assessed by measurements and hydrological models.

An **integral heat balance** within the domain D can be formulated in a similar manner:

$$\frac{dE}{dt} = \sum_i^{M_{E\_in}} J_{i\_in}(t) + \sum_i^{M_{E\_out}} J_{i\_out}(t) \tag{2.120}$$

The balance equation relates the total energy $E(t)$ within the domain D to the heat fluxes $J_{i\_in}(t)$ and $J_{i\_out}(t)$ (W) in and out of the domain. M_{E_in} and M_{E_out} are the number of the inflowing and outflowing heat flow components, respectively. Schematically, the heat fluxes are included in Figure 2.14. The inflow and outflow rates $J_i(t)$ comprise, in principle, advective heat flow rates based on the prevailing water flow rates, supplemented by conductive heat flow rates. The **heat flow rates** can be listed as follows:

- Advective heat inflow rate $J_{sw\_in}(t)$ from surface water bodies, like river and lakes

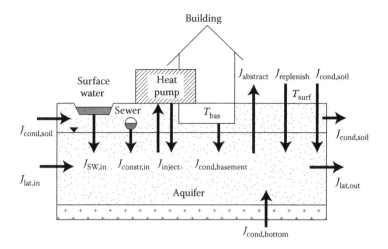

Figure 2.14 Thermal inflow and outflow rates for a schematic unconfined aquifer.

- Advective heat outflow rate $J_{sw\_out}(t)$ into surface water bodies
- Advective heat inflow rate $J_{replenish}(t)$ from distributed natural replenishment at soil surface or net loss due to areal evaporation or transpiration
- Advective lateral heat inflow rate $J_{lat\_in}(t)$ from upstream regions, or through lateral boundaries and the aquifer bottom into the saturated or unsaturated zones of domain D
- Advective lateral heat outflow rate $J_{lat\_out}(t)$ into downstream regions, or through lateral boundaries and the aquifer bottom into the saturated or unsaturated zones of domain D
- Advective heat inflow $J_{constr\_in}(t)$ from technical constructions (e.g., sewers, or infiltration of roof water, or from parking lots)
- Advective heat outflow $J_{constr\_out}(t)$ into constructions, like sewers
- Advective heat injection (inflow) $J_{inject}(t)$ from injection of heat by thermal use
- Advective heat abstraction (outflow) $J_{abstract}(t)$ from injection of heat by thermal use
- Conductive heat flow (inflow) $J_{cond\_soil}(t)$ through the soil surface into the domain D
- Conductive heat flow (inflow) $J_{cond\_lateral}(t)$ through the lateral boundaries and the bottom of the aquifer, into the domain D
- Conductive heat flow (inflow) $J_{cond\_constr}(t)$ from technical constructions (like sewers, or pipelines) into the domain D
- Conductive heat flow (inflow) $J_{cond\_basement}(t)$ from the basement of buildings into the domain D
- Heat injection (inflow) by BHEs, $J_{BHE\_in}(t)$
- Heat extraction (outflow) by BHEs, $J_{BHE\_out}(t)$

The **long-term heat balance** for the domain D, neglecting changes in energy storage within the aquifer for large values of time t, reads

$$\sum_{i=1}^{M_{w\_in}} \overline{J_{i\_in}} - \sum_{i=1}^{M_{w\_out}} \overline{J_{i\_out}} = 0 \qquad (2.121)$$

where $\overline{J_i}$ are time-averaged rates, with

$$\overline{J_i} = \frac{1}{t-t_0} \int_{t_0}^{t} J_i(t)\,dt \qquad (2.122)$$

The heat flow rates $J_i(t)$ or directly $\overline{J_i}$ have to be calculated or estimated, again, using measurements and models, and the water flow rates are as determined above. The question of reference temperature T_0 may be raised in this context. One possible reference temperature T_0 may be the average temperature $T_{surface}$ at the soil surface.

2.1.3.1 Rough estimation of the potential of an unconfined shallow aquifer for thermal use

A **simplified heat balance** can be obtained by considering long-term changes relative to a reference temperature T_0, and considering mean temperatures T_{gw} in shallow aquifers (saturated zone). Choosing the average temperature $T_{surface}$ at the soil surface as reference yields a relative temperature $T'(\mathbf{x}, t) = T(\mathbf{x}, t) - T_{surface}$, where T is the aquifer temperature. We further assume that the average temperature of surface water bodies is at the same reference temperature. With the relative temperature $T' = 0$ at the soil surface, and in the surface water bodies, the advective heat flow components J_{sw_in} and $J_{replenish}$, and possibly also the lateral component J_{lat_in}, may vanish depending on the thermal conditions. Furthermore, if the geothermal heat flux, as well as lateral heat advection, is neglected and no thermal use of the aquifer takes place, the groundwater temperature T_{gw} will also be at the relative temperature $T' = 0$. Now, we increase T_{gw} by ΔT due to thermal use, for example, due to the injection of warm water. A decrease ($\Delta T < 0$), on the other hand, will be obtained by the injection of cold water. For an increase of ΔT by the net heat fluxes from thermal use of groundwater J_{inject}, $J_{abstract}$, and J_{BHE}, the steady-state heat balance, neglecting conductive lateral heat fluxes, is

$$J_{sw\_out} + J_{lat\_out} + J_{constr\_out} + J_{cond\_soil} + J_{cond\_lat}$$
$$+ J_{cond\_constr} + J_{cond\_basement} + J_{thermal\_use} = 0 \qquad (2.123)$$

Again, note that **inflow is positive** and **outflow is negative**.

The **potential heat flux** J_{pot} for optimal thermal use can be expressed by

$$J_{pot} = J_{thermal\_use} \qquad (2.124)$$

The outflowing heat flux from surface water bodies J_{sw_out}, as well as J_{lat_out}, can be calculated based on the water outflow rate, for example, by modeling. Note that the sign of these rates depends on the sign of ΔT. For example, for $\Delta T < 0$, the advective heat outflow rate is negative. Heat flow from outflow into constructions J_{constr_out} or J_{cond_constr} might be small and can be roughly estimated. Lateral conductive heat transport J_{cond_lat} may diminish with time due to temperature equilibrium with the neighborhood of the aquifer. Vertical conductive heat flux $\overline{J_{cond\_soil}}$ from the soil surface to the aquifer (see Figures 2.13 and 2.14) can be roughly estimated by

$$J_{cond\_soil} = A_{soil} \lambda_{eff} \frac{T_{surface} - T_{gw}}{\left(f + \dfrac{m}{2} \right)_{soil}} = A_{soil} \lambda_{eff} \frac{-\Delta T}{\left(f + \dfrac{m}{2} \right)_{soil}} \tag{2.125}$$

where λ_{eff} is the effective thermal conductivity for the section between soil surface and the mean level of the saturated zone, and A_{soil} is the surface area of the soil within the aquifer. It includes sealed surfaces. In a similar manner, the vertical conductive heat flux $\overline{J_{cond\_basement}}$ from the basements of buildings to the aquifer can be approximated by

$$J_{cond\_basement} = A_{basement} \lambda_{vert} \frac{T_{basement} - T_{gw}}{\left(f + \dfrac{m}{2} \right)_{basement}} \tag{2.126}$$

where $A_{basement}$ is the integral area of basements from buildings.

Hötzl and Makurat (1981) and Menberg et al. (2013b) presented an example for a regional heat balance of an urban environment.

Consider an **illustrative, simple example** with a total area of 17 km², soil area A_{soil} = 14 km², mean distance of the soil from the aquifer $d = f + m/2$ = 12 m, and a total outflow rate of 0.7 m³ s⁻¹. Vertical conductive heat flux J_{cond_soil} from soil surface to the aquifer gets 7.0 MW, using an effective thermal conductivity of 2.0 W m⁻¹ K⁻¹ and a given temperature reduction of ΔT = −3 K. The vertical conductive heat flux from basements, with mean $d_{-basement} = f + m/2$ = 10 m, $A_{basement}$ = 3 km², and $T_{basement} - T_{gw}$ = 6 K, is 3.6 MW. The heat outflow rate J_{sw_out} is −8.8 MW. Neglecting further heat fluxes results in a potential heat flux for maximum thermal use on the order of 19 MW. Of course, the energy potential provided by the reduction of the groundwater temperature, which can be accomplished only once, is additive (about 1700 TJ). A similar calculation can be performed for $\Delta T > 0$ (injection of warm water).

Keep in mind that the performance J_{pot} is a theoretical value, which is difficult to achieve in practice because of the interaction by many single thermal installations. More realistically, we may introduce an **energy utilization factor**, which denotes the fraction of the theoretical heat flux that is feasible to harness. This utilization factor is typically expected to be smaller than 0.5 (Zhu et al. 2010). Nevertheless, the potential heat flux J_{pot} provides an **upper limit of the long-term potential for thermal use.** Preferably the heat flux calculations are performed sector-wise to take into account mainly the variability of f and m within the total area.

2.2 THERMAL PROPERTY VALUES

2.2.1 Heat capacity and thermal conductivity values

The **specific heat capacity** c of a material (pure material like water or solid material) is defined as its increase (or decrease) in heat (energy) due to a unit increase (or decrease) in temperature, related to unit mass. Commonly used units are J kg^{-1} K^{-1} or W s kg^{-1} K^{-1}. An overview on the specific heat capacity of various minerals can be found in Waples and Waples (2004a,b) and Clauser (2011a). It can also be expressed as **volumetric heat capacity** C (also for mixtures, like soils) with commonly used units J m^{-3} K^{-1} or W s m^{-3} K^{-1}. The volumetric heat capacity C_m for soils or aquifers with porosity ϕ and water saturation S_w can be calculated using Equation 2.46. Similarly, the effective heat capacity of material with multiple components such as different minerals in soil is obtained by the volume-weighted mean of the individual fractions. Typically used values for thermal capacity of water, pure materials, and porous media from the literature are shown in Tables 2.1 through 2.6.

Table 2.2 Density, specific heat capacity, and thermal conductivity of water

Temperature (°C)	Density[a] ρ_w (kg m^{-3})	Specific heat capacity[a] c_w (J kg^{-1} K^{-1})	Thermal conductivity[b] λ_w (W m^{-1} K^{-1})
0.1	999.84	4217.0	
5	999.97		0.5675
10	999.70	4190.6	0.5781
15	999.10		0.5881
20	998.21	4156.7	0.5975
25	997.05	4137.6	0.6064
30	995.65	4117.2	0.6147

[a] CRC 2011.
[b] Ramires et al. 1995.

Table 2.3 Volumetric heat capacity and thermal conductivity of soil for various water contents

Soil	θ_w (m^3 m^{-3})	λ_m (W m^{-1} K^{-1})	C_m (J m^{-3} K^{-1})
Sandy soil ($\phi = 0.4$)	0	0.3	1.28×10^6
	0.2	1.8	2.12×10^6
	0.4	2.2	2.96×10^6
Clay soil ($\phi = 0.4$)	0	0.25	2×10^6
	0.2	1.18	3.10×10^6
	0.4	1.58	5.76×10^6

Source: After Williams, P.J. and Smith, M.W., *The Frozen Earth. Fundamentals of Geocryology.* Cambridge University Press, Cambridge, UK, 1989.

Table 2.4 Density, volumetric heat capacity, and thermal conductivity range of common aquifers and building materials

Material	ρ_m (kg m^{-3})	λ_m (W m^{-1} K^{-1})	C_m (J m^{-3} K^{-1})
Clay, silt, dry	1800–2000	0.4–1.0	1.5×10^6–1.6×10^6
Clay, silt, saturated	2000–2200	0.9–2.3	2.0×10^6–2.8×10^6
Sand, dry	1800–2200	0.3–0.8	1.3×10^6–1.6×10^6
Sand, saturated	1900–2300	1.5–4.0	2.2×10^6–2.8×10^6
Gravel, blocks, dry	1800–2200	0.4–0.5	1.3×10^6–1.6×10^6
Gravel, blocks, sat.	1900–2300	1.6–2.0	2.2×10^6–2.6×10^6
Clay, siltstone	2400–2600	1.1–3.5	2.1×10^6–2.4×10^6
Sandstone	2200–2700	1.3–5.1	1.8×10^6–2.6×10^6
Marble	2300–2600	1.5–3.5	2.2×10^6–2.3×10^6
Limestone	2400–2700	2.5–4.0	2.1×10^6–2.4×10^6
Dolomite	2400–2700	2.8–4.3	2.1×10^6–2.4×10^6
Granite	2400–3000	2.1–4.1	2.1×10^6–3.0×10^6
Bentonite		0.5–0.8	$\cong 3.9 \times 10^6$
Concrete	$\cong 2000$	0.9–2.0	$\cong 1.8 \times 10^6$
Steel	7800	60	3.12×10^6

Source: After VDI-Richtlinie 4640. Thermische Nutzung des Untergrundes (Guideline for thermal use of the underground). Verein Deutscher Ingenieure, VDI-Gesellschaft Energietechnik, Germany, 2012.

According to Fourier, **thermal conductivity** λ is the coefficient in the heat conduction equation. It expresses the ability of a material (pure material like water or solid material, or a mixture as present in soils) to conduct heat. The commonly used unit is W m^{-1} K^{-1}. Typically used values for the thermal conductivity of water, pure materials, and porous media from the literature are shown in Tables 2.1 through 2.6. An overview on the thermal conductivity of various rock material and minerals can be found in Waples and Waples (2004a,b), Clauser (2011b), and Banks (2008). Various models exist to express the heat thermal conductivity of soils or aquifers (Section 2.1.2.2). At this stage, we would like to recall the **arithmetic mean model** (Equation 2.54), the **harmonic mean model** (Equation 2.55), and the **geometric mean model** (Equation 2.67) for the thermal conductivity for saturated aquifer material with porosity ϕ. The coded functions (MATLAB scripts) of Equations 2.54, 2.55, and 2.67 can be found at http://www.crcpress.com/product/isbn/9781466560192 under the name "Tcond_arithmetic.m," "Tcond_harmonic.m," and "Tcond_geometric.m," respectively. A comparison of the results from the three models for a chosen porosity of $\phi = 0.25$ is depicted in Figure 2.15. It shows that, as expected, the arithmetic model represents the upper values and the harmonic model represents the lower ones, while the geometric model lies in between.

Table 2.5 Specific heat capacity and density of different soil-forming minerals and calculated volumetric heat capacities

Mineral groups	Contents of groups	Density of Minerals ρ_s (kg m⁻³)[a]	Arithmetic average of density ρ_s (kg m⁻³)	Specific heat capacity of minerals c_s (MJ kg⁻¹ K⁻¹)[b] at 25°C, 0.74[c]	Arithmetic average of specific heat capacity c_s (MJ kg⁻¹ K⁻¹)	Arithmetic mean of the volumetric heat capacity C_s of mixed minerals (J m⁻³ K⁻¹)
Tectosilicates Silica group	α-Quartz (trigonal) SiO₂	2650	2650	0.74	0.74	2 × 10⁶
Tectosilicates Feldspar group	Albite NaAlSi₃O₈	2650	2600	0.71	0.77	2 × 10⁶
	Anorthite CaAl₂SiO₈	2750		0.73		
	Oligoclase (Na, Ca)(Si, Al)₄O₈	2650		0.85		
	Orthoclase KAlSi₃O₈	2600		0.70		
	Microcline KAlSi₃O₈	2600		0.70		
Orthosilicates Olivine group	Fayalite Fe₂SiO₄	3400	3300	0.79	0.79	2.6 × 10⁶
	Forsterite Mg₂SiO₄	3300		0.79		
	Monticellite CaMgSiO₄	3300		0.80		
Oxide group	Corundum Al₂O₃	4000	4800	0.71	0.71	3.4 × 10⁶
	Hematite Fe₂O₃	5300		0.72		
	Magnetite Fe₃O₄	5100		0.70		
	Ilmenite FeTiO₃	4700		0.70		

(continued)

Table 2.5 (Continued) Specific heat capacity and density of different soil-forming minerals and calculated volumetric heat capacities

Mineral groups	Contents of groups	Density of Minerals ρ_s (kg m^{-3})[a]	Arithmetic average of density ρ_s (kg m^{-3})	Specific heat capacity of minerals c_s (MJ kg^{-1} K^{-1})[b]	Arithmetic average of specific heat capacity c_s (MJ kg^{-1} K^{-1})	Arithmetic mean of the volumetric heat capacity C_s of mixed minerals (J m^{-3} K^{-1})
Phyllosilicates Mica group (clay minerals)	Chlorite (Mg, Fe)$_3$ [(Si, Al)$_4$O$_{10}$[OH]$_2$]	2800	2900	0.93	0.90	2.6 × 10^6
	Kaolinite	2700		0.95		
	Serpentine (Mg, Fe)$_3$[Si$_2$O$_5$][OH]$_4$	2900		1.00		
	Pyrophyllite Al$_2$Si$_4$O$_{10}$[OH]$_2$	2900		0.90		
	Biotite K(Mg, Fe)$_3$ (Al, Fe^{+3})Si$_3$O$_{10}$(OH, F)$_2$	3000		0.80		
	Muscovite KAl$_2$[Si$_3$Al] O$_{10}$[OH]$_2$	2800		0.82		
Chain silicates Amphibole group	Actinolite Ca$_2$ (Mg, Fe)$_5$[Si$_8$O$_{22}$][OH]$_2$	3100	3300	0.80	0.80	2.6 × 10^6
	Hornblende Ca$_2$(Mg, Fe)$_4$(Al, Fe^{+3}) [Si$_7$Al]O$_{22}$[OH]$_2$	3300		0.84		
	Pargasite NaCa$_2$(Mg, Fe)$_4$Al[Si$_6$Al$_2$]O$_{22}$[OH]$_2$	3100		0.78		
	Rhodonite (Mn^{2+}, Fe^{2+}, Mg, Ca)SiO$_3$	3600		0.75		

Chain silicates Pyroxene group	Augite (Ca, Na)(Mg, Fe, Al, Ti)(Al, Si)$_2$O$_6$	3400	0.75	3300	0.77	2.6×10^6
	Diopside CaMgSi$_2$O$_6$	3400	0.77			
	Enstatite Mg$_2$Si$_2$O$_6$	3200	0.77			
	Jadeite Na(Al, Fe)Si$_2$O$_6$	3300	0.78			
Nonsilicates Sulfide group	Chalcopyrite CuFeS$_2$	4200	0.53	4800	0.45	2.2×10^6
	Sphalerite (Zn, Fe^{2+})S	4100	0.43			
	Arsenopyrite FeAsS	6000	0.43			
Nonsilicates Carbonate group	Calcite (trigonal) CaCO$_3$	2700	0.80	2750	0.88	2.4×10^6
	Dolomite CaMg[CO$_3$]$_2$	2800	0.91			

Source: Erol, S., Estimation of heat extraction rates of GSHP systems under different hydrogeological conditions, MSc. thesis, University of Tübingen, 85 pp., 2011.

Note: Thermal conductivity values are taken in a temperature range between 10°C and 35°C.

[a] Webmineral 2011.
[b] Clauser 2006.
[c] Grønvold et al. 1989.

Table 2.6 Calculated thermal properties of dry soils and rocks based on the mixed minerals

Mineral contents (%)[a]	Selected soils and rocks	Region[b]	Volumetric heat capacity C_m of soils and rocks. Arithmetic average (MJ m⁻³ K⁻¹)	Thermal conductivity λ_m of soils and rocks. Geometric average (W m⁻¹ K⁻¹)
65% quartz, 20% potassium feldspar, 12% sodium feldspar, 1% muscovite, 1% biotite, 1% hornblende	Gneiss	Southeast Germany (Bayern)	2.1	4.4
35% K-feldspar, 30% plagioclase, 29% quartz, 5% biotite, 1% amphibole	Granite		2.0	3.2
43% plagioclase, 41% pyroxene, 5% alkali feldspar, 5% silica class, 5% olivine, 1% magnetite and ilmenite	Basalt		2.3	3.0
100% calcium and magnesium carbonates	Dolomite	Southwest Germany (Baden-Württemberg)	2.6	5.2[c]
100% calcite	Limestone		2.2	3.1[d]
50% quartz, 35% calcite, 13% plagioclase, 1% K-feldspar, 1% alkali feldspar	Gravel		2.0	4.5
90% quartz, 5% clay minerals, 5% K-feldspar	Sandstone	Middle Germany (Hessen)	2.0	5.5
90% clay minerals, 5% quartz, 3% carbonate, 1% feldspar, 1% sulfide	Clay	Northeast Germany (Brandenburg)	2.5	3.2
98% quartz, 1% feldspar, 1% clay minerals	Sand	North Germany (Mecklenburg Vorpommern)	2.0	6.0
60% quartz, 40% clay minerals	Silt	West Germany (Nordrhein-Westfalen)	2.2	4.6

Source: Erol, S., Estimation of heat extraction rates of GSHP systems under different hydrogeological conditions, MSc. thesis, University of Tübingen, 85 pp., 2011.

Note: The thermal conductivity and volumetric heat capacity of soils and rocks are taken for dry conditions.

a Press and Siever 1998.
b BGR 2011.
c VDI 2010.
d Thomas Jr. et al. 1973.

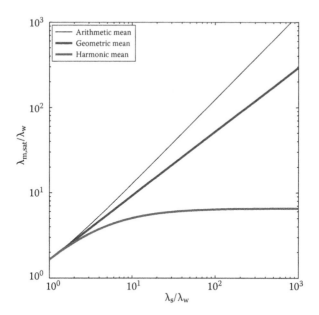

Figure 2.15 Comparison of the results from the three models "Tcond_arithmetic.m," "Tcond_harmonic.m," and "Tcond_geometric.m." (Modified after Woodside, W. and Messmer, J.H., *Journal of Applied Physics* 32 (9), 1688–1698, 1961.)

Markle et al. (2006) characterized thermal conductivity of a sand and gravel aquifer in Ontario, Canada, in detailed spatial resolution. They determined thermal conductivity values in a vertical two-dimensional profile by (1) measuring the thermal conductivity of solids and the mineral composition, (2) measuring the volumetric water content using cross-hole ground-penetrating radar, (3) evaluating several models for the effective thermal conductivity, (4) calculating the distribution using the selected model, and (5) simulating the thermal transport. The apparent thermal conductivity λ_m ranged between 2.14 and 2.69 W m^{-1} K^{-1} with a mean of 2.42 W m^{-1} K^{-1}. They found that the heterogeneous thermal conductivity field results in an increased thermal dispersion, which is most pronounced at the thermal front.

REFERENCES

Allen, R.G., Pereira, L.S., Raes, D., Smith, M. (2006). Crop evapotranspiration. Guidelines for computing crop water requirements. FAO Irrigation and Drainage Paper No. 56.

Balland, V., Arp, P.A. (2005). Modeling soil thermal conductivities over a wide range of conditions. *Journal of Environmental Engineering and Science* 4, 549–558, doi:10.1139/S05-007.

Banks, D. (2008). *An Introduction to Thermogeology: Ground Source Heating and Cooling.* Blackwell Publishing, Oxford, UK.

Bauer, D., Heidemann, W., Müller-Steinhagen, H., Diersch, H.-J.G. (2011). Thermal resistance and capacity models for borehole heat exchangers. *International Journal of Energy Research* 35, 312–320.

Bayer, P., Huggenberger, P., Renard, P., Comunian, A. (2011). Three-dimensional high resolution fluvio-glacial aquifer analog, Part 1: Field study. *Journal of Hydrology* 405, 1–9.

Bear, J. (1972). *Dynamics of Fluids in Porous Media.* Elsevier, New York, USA.

Bear, J. (1979). *Hydraulics of Groundwater.* McGraw-Hill, New York, USA.

Bear, J., Cheng, A.H.-D. (2010). *Modelling Groundwater Flow and Contaminant Transport.* Springer, Dordrecht, the Netherlands.

BGR (2011). Bundesanstalt für Geowissenschaften und Rohstoffe. Accessed on July 20, 2011, at http://www.bgr.bund.de.

Brooks, R.H., Corey, A.T. (1966). Properties of porous media affecting fluid flow. *Journal of the Irrigation and Drainage Division, Proceedings of the American Society of Civil Engineers ASCE* 92(2), 61–88.

Campbell, G.S., Jungbauer, J.D., Bidlake, W.R., Hungerford, R.D. (1994). Predicting the effect of temperature on soil thermal conductivity. *Soil Science* 158(5), 307–313.

Carslaw, H.S., Jaeger, J.C. (1959). *Conduction of Heat in Solids.* Oxford University Press, Oxford, UK.

Chang, C.-M., Yeh, H.-D. (2012). Stochastic analysis of field-scale heat advection in heterogeneous aquifers. *Hydrological Earth System Science HESS* 16, 641–648.

Chen, S.X. (2008). Thermal conductivity of sands. *International Journal of Heat and Mass Transfer* 44, 1241–1246.

Claesson, J., Hellström, G. (2011). Multipole method to calculate borehole thermal resistance in a borehole heat exchanger. *HVAC&R Research* 17(6), 895–911.

Clauser, C. (Ed.) (2003). *Numerical Simulation of Reactive Flow in Hot Aquifers—SHEMAT and Processing SHEMAT.* Springer, Berlin.

Clauser, C. (2006). Geothermal energy. In: K. Heinloth (Ed.), *Landolt-Börnstein—Numerical Data and Functional Relationships.* Springer Verlag, Heidelberg-Berlin.

Clauser, C. (2011a). Thermal storage and transport properties of rocks, I: Heat capacity and latent heat. In: H.K. Gupta (Ed.), *Encyclopedia of Solid Earth Geophysics.* 2nd ed., Springer, Dordrecht, pp. 1423–1431.

Clauser, C. (2011b). Thermal storage and transport properties of rocks, II: Thermal conductivity and diffusivity. In: H.K. Gupta (Ed.), *Encyclopedia of Solid Earth Geophysics.* 2nd ed., Springer, Dordrecht, pp. 1431–1448.

Constantz, J. (2008). Heat as a tracer to determine streambed water exchanges. *Water Resources Research* 44, W00D10, doi:10.1029/2008WR006996.

CRC (2011). *Handbook of Chemistry and Physics.* W.M. Haynes (Ed.), CRC Press, Boca Raton, FL, USA.

Dagan, G. (1989). *Flow and Transport in Porous Formations.* Springer, Berlin.

de Marsily, G. (1986). *Quantitavive Hydrogeology.* Academic Press, Orlando, USA.

Dentz, M., Carrera, J. (2005). Effective solute transport in temporally fluctuating flow through heterogeneous porous media. *Water Resources Research* 41, W08414, doi:10.1029/2004WR003571.

de Vries, D.A. (1963). Thermal properties of soil. In: W.R. van Wijk (Ed.), *Physics of Plant Environment*. North-Holland, Amsterdam, pp. 210–235.

Diersch, H.-J.G., Bauer, D., Heidemann, W., Rühaak, W., Schätzl, P. (2011). Finite element modeling of borehole heat exchanger systems. Part 1. Fundamentals. *Computer and Geosciences* 37, 122–1137.

Domenico, P.A., Palciauskas, V.V. (1973). Theoretical analysis of forced convective heat transfer in regional groundwater flow. *Geological Society of America Bulletin* 84(12), 3803–3814.

Erol, S. (2011). Estimation of heat extraction rates of GSHP systems under different hydrogeological conditions. MSc. thesis, University of Tübingen, Tübingen, Germany, 85 pp.

Eskilson, P., Claesson, J. (1988). Simulation model for thermally interacting heat extraction boreholes. *Numerical Heat Transfer* 13(2), 149–165.

Farouki, O.T. (1981). The thermal properties of soils in cold regions. *Cold Regions Science and Technology* 5, 67–75.

Ferguson, G. (2007). Heterogeneity and thermal modeling of ground water. *Ground Water* 45(4), 458–490.

Ferguson, G., Beltrami, H., Woodbury, A. (2006). Perturbation of ground surface temperature reconstructions by groundwater flow? *Geophysical Research Letters* 33, L13708, doi:10.1029/2006GL026634.

Fujii, H., Itoi, R., Fujii, J., Uchida, Y. (2005). Optimizing the design of large-scale ground-coupled heat pump systems using groundwater and heat transport modeling. *Geothermics* 34(3), 347–364.

Geiger, S., Emmanuel, S. (2010). Non-Fourier thermal transport in fractured geological media. *Water Resources Research* 46, W07504, doi:10.1029/2009WR008671.

Gelhar, L.W. (1993). *Stochastic Subsurface Hydrology*. Englewood Cliffs, New Jersey, Prentice-Hall.

Gelhar, L.W., Welty, C., Rehfeldt, K.R. (1992). A critical review of data on field-scale dispersion in aquifer. *Water Resources Research* 28(7), 1955–1974.

Giakoumakis, S.G. (1994). A model for predicting coupled heat and mass transfers in unsaturated partially frozen soil. *International Journal of Heat and Fluid Flow* 15(2), 163–171.

Green, D.W., Perry, R.H., Babcock, R.E. (1964). Longitudinal dispersion of thermal energy through porous media with a flowing fluid. *AIChE Journal* 10(5), 645–651.

Gröber, H., Erk, S., Grigull, U. (1955). *Grundgesetze der Wärmeübertragung*. U. Grigull (rev. ed.), Springer, Berlin, Germany.

Grønvold, F., Stølen, S., Svendsen, S.R. (1989). Heat capacity of [alpha] quartz from 298.15 to 847.3 K, and of [beta] quartz from 847.3 to 1000 K-transition behaviour and revaluation of the thermodynamic. *Thermochimica Acta* 139, 225–243.

Hansson, K., Šimůnek, J., Mizoguchi, M., Lundin, L.C., van Genuchten, M.T. (2004). Water and heat transport in frozen soils: Numerical solutions and freeze-thaw applications. *Vadose Zone Journal* 3, 693–704.

Hellström, G. (1991). Ground heat storage. Thermal analyses of duct storage systems. PhD Thesis, Lund University, Lund, Sweden.

Hidalgo, J.J., Carrera, J., Dentz, M. (2009). Steady state heat transport in 3D heterogeneous porous media. *Advances in Water Resources* 32(8), 1206–1212.

Hopmans, J.W., Simunek, J., Bristow, K.L. (2002). Indirect estimation of soil thermal properties and water flux using heat pulse probe measurements: Geometry and dispersion effects. *Water Resources Research* 38(1) 1006, 7-1–7-14, doi: 10.1029/2000WR000071.

Hötzl, H., Makurat, A. (1981). Veränderungen der Grundwassertemperaturen unter dicht bebauten Flächen am Beispiel der Stadt Karlsruhe. *Zeitschrift der deutschen geologischen Gesellschaft* 132, 767–777.

Hsu, C.T., Cheng, P. (1990). Thermal dispersion in a porous medium. *International Journal of Heat and Mass Transfer* 33(8), 1587–1597.

Incropera, F.P., Dewitt, D.P., Bergman, T.L., Lavine, A.S. (2007). *Introduction to Heat Transfer*. 5th ed., J. Wiley & Sons, Hoboken, NJ, USA.

Johansen, O. (1975). Thermal conductivity of soils. PhD thesis Univ. of Trondheim (Translation US Army Cold Regions Research and Engineering Laboratory, Hanover, NH, USA, 1977).

Jun, L., Xu, Z., Jun, G., Jie, Y. (2009). Evaluation of heat exchange rate of GHE in geothermal heat pump systems. *Renewable Energy* 34, 2898–2904.

Jussel, P., Stauffer, F., Dracos, T. (1994). Transport modeling in heterogeneous aquifer: 1. Statistical description and numerical generation of gravel deposits. *Water Resources Research* 30(6), 1803–1817.

Kersten, M.S. (1949). Final report, laboratory research for the determination of the thermal properties of soils. Corps of Engineers, US Army, Univ. Minnesota Engineering Experiment Station, Bulletin No. 28.

Kinzelbach, W., Ackerer, P. (1986). Modélisation de la proagation d'un contaminant dans un champ d'écoulement transitoire. *Hydrogéolgie* 2, 197–206.

Kollet, S.J., Cvijanovic, I., Schüttemeyer, D., Maxwell, R.M., Moene, A.F., Bayer P. (2009). The influence of rain sensible heat and subsurface energy transport on the energy balance at the land surface. *Vadose Zone Journal* 8(4), 846–857.

Kunii, D., Smith, J.M. (1960). Heat transfer characterristics of porous rocks. *AIChE Journal* 6(1), 71–78.

Lamarche, L., Kajl, S., Beauchamp, B. (2010). A review of methods to evaluate borehole thermal resistances in geothermal heat pump systems. *Geothermics* 39, 187–200.

Levec, J., Carbonell, R.G. (1985). Longitudinal and lateral thermal dispersion in packed beds. Part II. Comparison between theory and experiment. *AIChE Journal* 31(4), 591–602.

Lo Russo, S., Taddia, G. (2010). Advective heat transport in an unconfined aquifer induced by field injection of an open-loop groundwater heat pump. *American Journal of Environmental Sciences* 6(3), 253–259.

Lu, X. (2009). Experimental investigation of thermal dispersion in saturated soils with one-dimensional water flow. *Soil Science Society of America Journal* 73(6), 1912–1920.

Ma, R., Zheng, C. (2010). Effects of density and viscosity in modeling heat as a groundwater tracer. *Ground Water* 48(3), 380–389.

Marcotte, D., Pasquier, P. (2008). On the estimation of thermal resistance in borehole thermal conductivity tests. *Renewable Energy* 33, 2407–2415.

Markle, J.M., Schincarion, R.A., Sass, J.H., Molson, J.M. (2006). Characterizing the two-dimensional thermal conductivity distribution in a sand and gravel aquifer. *Soil Science Society of America Journal* 70, 1281–1294.

Menberg, K., Steger, H., Zorn, R., Reuß, M., Pröll, M., Bayer, P., Blum, P. (2013a). Bestimmung der Wärmeleitfähigkeit im Untergrund durch Labor- und Feldversuche und anhand theoretischer Modelle. *Grundwasser* 18(2), 103–116.

Menberg, K., Blum, P., Schaffitel, A., Bayer, P. (2013b). Long-term evolution of anthropogenic heat fluxes into asubsurface urban heat island. *Environmental Science and Technology*, in press.

Mercer, J.W., Faust, C.R., Miller, W.J., Pearson, F.J., Jr. (1982). Review of simulation techniques for aquifers thermal energy storage (ATES). In: V.T. Chow (Ed.), *Advances in Hydroscience* 13, 1–129. Academic Press, New York, USA.

Metzger, T., Didierjean, S., Maillet, D. (2004). Optimal experimental estimation of thermal dispersion coefficients in porous media. *International Journal of Heat and Mass Transfer* 47(14–16), 3341–3353.

Molina-Giraldo, N., Bayer, P., Blum, P. (2011). Evaluating the influence of thermal dispersion on temperature plumes from geothermal systems using analytical solutions. *International Journal of Thermal Sciences* 50, 1223–1231.

Molson, J.W., Frind, E., Palmer, C.D. (1992). Thermal energy storage in an unconfined aquifer: 2. Model development, validation, and application. *Water Resources Research* 28(10), 2857–2867.

Moyne, C., Didierjean, S., Amaral Souto, H.P., da Silveira, O.T. (2000). Thermal dispersion in porous media: One-equation model. *International Journal of Heat and Mass Transfer* 43(20), 3853–3867.

Neuman, S.P. (1990). Universal scaling of hydraulic conductivities and dispersivities in geologic media. *Water Resources Research* 26(8), 1749–1758.

Nield, D.A., Bejan, A. (2006). *Convection in Porous Media*. Springer, New York, USA.

Oostrom, M., Hayworth, J.S., Dane, J.H., Güven, O. (1992). Behavior of dense aqueous phase leachate plumes in homogeneous porous media. *Water Resources Research* 28(8), 2123–2134.

Parlange, M.B., Cahill, A.T., Nielsen, D.R., Hopmans, J.W., Wendroth, O. (1998). Review of heat and water movement in field soils. *Soil and Till Research* 47, 5–10.

Pedras, M.H.J., de Lemos, M.J.S. (2008). Thermal dispersion in porous media as a function of the solid-fluid conductivity ratio. *International Journal of Heat and Mass Transfer* 51(21–22), 5359–5367.

Press, F., Siever, R. (1998). *Understanding Earth*. 2nd ed., W.H. Freeman, New York.

Ramires, M.L.V., Nieto de Castro, C.A., Nagasaka, Y., Nagashima, A., Asseql, M.J., Wakeham, W.A. (1995). Standard reference data for the thermal conductivity of water. *Journal of Physical and Chemical Reference Data* 24(3), 1377–1381.

Rau, G., Andersen, M.S., Acworth, R.I. (2012). Experimental investigation of the thermal dispersivity term and its significance in the heat transport equation for flow in sediments. *Water Resources Research* 48, W03511, doi:10.1029/2011WR011038.

Reiter, M. (2001). Using precision temperature logs to estimate horizontal and vertical groundwater flow components. *Water Resources Research* 37(3), 663–674.

Sagia, Z., Stegou, A., Rakopoulos, C. (2012). Borehole resistance and heat conduction around vertical ground heat exchangers. *The Open Chemical Engineering Journal* 6, 32–40.

Sauty, J.P., Gringarten, A.C., Fabris, H., Thiery, D., Menjoz, A., Landel, P.A. (1982). Sensible energy storage in aquifers 2. Field experiments and comparison with theoretical results. *Water Resources Research* 18(2), 253–265.

Schaap, M.G., van Genuchten, M.T. (2006). A modified Mualem–van Genuchten formulation for improved description of the hydraulic conductivity near saturation. *Vadose Zone Journal* 5, 27–34.

Schulze-Makuch, D. (2005). Longitudinal dispersivity data and implications for scaling behaviour. *Ground Water* 43(3), 443–456.

Sharqawy, M.H., Morkheimer, E.M., Bader, H.M. (2009). Effective pipe-to-borehole thermal resistance for vertical ground heat exchangers. *Geothermics* 38, 271–277.

Shook, M. (2001). Predicting thermal breakthrough in heterogeneous media from tracer tests. *Geothermics* 30, 573–589.

Smith, L., Chapman, D.S. (1983). Thermal effects of groundwater flow. 1. Regional scale systems. *Journal of Geophysical Research* 88(B1), 593–608.

Sutton, M.G., Nutter, D.W., Couvillion, R.J. (2003). A ground resistance for vertical bore heat exchangers with groundwater flow. *Journal of Energy Resources Technology ASME* 125(3), 183–189.

Taniguchi, M., Shimada, J., Tanaka, T., Kayane, I., Sakura, Y., Shimano, Y., Dapaah-Siakwan, S., Kawashima, S. (1999). Disturbances of temperature-depth profiles due to surface climate change and subsurface water flow: 1. An effect of linear increase in surface temperature caused by global warming and urbanization in the Tokyo metropolitan area, Japan. *Water Resources Research* 35(5), 1507–1517.

Tarnawski, V.R., Wagner, B. (1993). Modeling the thermal conductivity of frozen soils. *Cold Regions Science and Technology* 22, 19–31.

Thomas, J., Jr., Frost, R.R., Harvey, R.D. (1973). Thermal conductivity of carbonate rocks. *Engineering Geology* 7(1), 3–12.

van Genuchten, R. (1980). A closed form equation for predicting the hydraulic conductivity of unsaturated soils. *Soil Science Society of America Journal* 44, 892–898.

Vandenbohede, A., Louwick, A., Lebbe, L. (2009). Conservative solute vs. heat transport in porous media during push-pull tests. *Transport in Porous Media* 76, 265–287.

VDI (2012). VDI-Richtlinie 4640: Thermische Nutzung des Untergrundes (Guideline for thermal use of the underground). Verein Deutscher Ingenieure, VDI-Gesellschaft Energietechnik, Germany.

Wagner, V., Blum, P., Kübert, M., Bayer, P. (2013). Analytical approach to groundwater-influenced thermal response tests of grouted borehole heat exchangers. *Geothermics* 46, 22–31.

Wang, H., Yang, B., Xie, J., Qi, C. (2012). Thermal performance of borehole heat exchangers in different aquifers: A case study from Shouguang. *International Journal of Low Carbon Technologies*, doi:10.1093/ijlct/cts043.

Waples, D.W., Waples, J.S. (2004a). A review and evaluation of specific heat capacities of rocks, minerals, and subsurface fluids. Part 1: Minerals and Nonporous rocks. *Natural Resources Research* 13(2), 97–122.

Waples, D.W., Waples, J.S. (2004b). A review and evaluation of specific heat capacities of rocks, minerals, and subsurface fluids. Part 2: Fluids and porous rocks. *Natural Resources Research* 13(2), 123–130.

Ward, J.D., Simmons, C.T., Dillon, P.J. (2007). A theoretical analysis of mixed convection in aquifer storage and recovery: How important are density effects? *Journal of Hydrology* 343, 169–186.

Webmineral. (2011). Webmineral: Basic information for minerals. Accessed on May, 15, 2011, at http://webmineral.com.

Whitaker, S. (1977). Simultaneous heat, mass and momentum transfer in porous media: A theory of drying. *Advances in Heat Transfer* 13, 119–203.

Williams, P.J., Smith, M.W. (1989). *The Frozen Earth. Fundamentals of Geocryology*. Cambridge University Press, Cambridge, UK.

Woodbury, A.D., Smith, L. (1985). On the thermal effects of three-dimensional groundwater flow. *Journal of Geophysical Research* 90(B1), 759–767.

Woodside, W., Messmer, J.H. (1961). Thermal conductivity of porous media. 1. Unconsolidated sands. *Journal of Applied Physics* 32(9), 1688–1698.

Woumeni, R.S., Vauclin, M. (2006). A field study of the coupled effects of aquifer stratification, fluid density, and groundwater fluctuations on dispersion assessments. *Advances in Water Resources* 29, 1037–1055.

Yang, W., Shi, M., Liu, G., Chen, Z. (2009). A two-region simulation model of vertical U-tube ground heat exchanger and its experimental verification. *Applied Energy* 86, 2005–2012.

Xu, M., Eckstein, Y. (1995). Use of weighted least-squares method in evaluating of the relationship between dispersivity and field scale. *Ground Water* 33(6), 905–908.

Zarrella, A., Scarpa, M., De Carli, M. (2011). Short time step analysis of vertical ground-coupled heat exchangers: The approach of CaRM. *Renewable Energy* 36, 2357–2367.

Zeng, H., Diao, N., Fang, Z. (2003). Heat transfer analysis of boreholes in vertical ground heat exchangers. *International Journal of Heat and Mass Transfer* 46, 4467–4481.

Zhu, K., Blum, P., Ferguson, G., Balke, K.-D., Bayer, P. (2010). Geothermal potential of urban heat islands. *Environmental Research Letters* 5, 044002, doi:10.1088/1748-9326/5/4/044002.

Chapter 3

Analytical solutions

In general, **flow and heat transport problems can only be solved analytically for special cases**. Such cases are characterized as follows:

- Constant coefficients of the flow and heat transport equation and of the boundary conditions and therefore homogeneous porous media are assumed. In the case of heat transport, this usually means constant thermal retardation or thermal velocity and constant thermal diffusion and dispersion coefficients.
- The flow domain is sufficiently simple, for example, infinite, or radially symmetric.
- The initial condition is sufficiently simple, for example, constant, or zero.
- The boundary conditions are sufficiently simple, for example, constant, or zero.

If these conditions are met to some degree, analytical solutions are, in general, preferable over numerical ones. Important applications of analytical solutions are **analytical approximations** to complex situations, the **determination of parameters** using experimental data, or the **test of numerical solution methods**, like the finite difference, the finite volume, or the finite element methods.

For heat transport, in general, analytical solutions are restricted to closed systems. Nevertheless, analytical approximations are available for both open- and closed-loop systems. There exists a long tradition in using **analytical solutions** for groundwater flow, and mass diffusion, solute transport, and heat transport problems in porous media and in fluids. In the following literature overview, we list a few representatives in the field of heat transport in solid materials, saturated porous media, and groundwater.

Carslaw and Jaeger (1946, 1959) presented a large number of analytical solutions to the problem of heat conduction in solid materials. Many of these solutions are used in the field of both water flow (exploiting the analogy to piezometric head diffusion) and heat transport in groundwater.

Ingersoll et al. (1948, 1954) presented a comprehensive theory and analytical solutions to the heat conduction problem with engineering and geological applications. Among other subjects, they treated the mathematical problem of heat sources for heat pumps.

Domenico and Palciauskas (1973) offer analytical solutions to the steady-state flow and heat transport problem in homogeneous rectangular vertical regions using concepts from Toth (1963). The upper boundary condition is a prescribed head condition that represents the water table. The lateral and lower boundaries are impermeable.

Gringarten and Sauty (1975) investigated the transient temperature evolution of a pumped aquifer during reinjection of water at a temperature different from that of the native water. A horizontal aquifer of constant thickness with impermeable bottom and top layers is considered. Flow is assumed to be at steady state, thus neglecting the short transient period during reinjection. Flow direction is arbitrary. Transient heat transport is solved semianalytically for the curved stream channel between the two wells, using the stream function concept and taking into account heat flow from cap rock and bedrock. Results are given in dimensionless form.

Mercer et al. (1982) reviewed a series of analytical solutions for aquifer thermal energy storage.

Uffink (1983) investigated the heat exchange between aquifer and adjacent aquitard layers. He developed simplified analytical solutions for advective–diffusive heat transport in an aquifer close to injection wells for transient and periodic conditions. Heat transport in the top and bottom layers is assumed to be vertical and conductive. For a thin aquifer, he adopted Carslaw and Jaeger's (1959) solution for one-dimensional, purely advective heat transport and exchange with adjacent semi-infinite layers. The further assumption for vertical temperature profiles in thin aquifers is referred to as Lauwerier's (1955) assumption. As already shown by Gringarten and Sauty (1975), the approach is also valid for two-dimensional heat transport if heat transport perpendicular to streamlines is neglected. Due to the heat exchange, considerable damping of temperature changes may take place. Uffink (1983) further shows that for periodic boundary conditions, the thickness of the aquifer (typically thicker than a few meters) has to be taken into account for the vertical heat exchange.

Güven et al. (1983) derived analytical expressions for the temperature distribution of a simplified aquifer thermal energy storage concept, taking heat exchange at the soil surface into account. The system is restricted to heat conduction processes in a cylindrical region.

A unified mathematical analysis and analytical solutions of heat and mass diffusion problems were presented by Mikhailov and Özişik (1984) and Häfner et al. (1992).

Bundschuh (1993) formulated analytical models for the simulation of periodic temperature variations in shallow aquifer.

Lu and Ge (1996) developed an analytical solution for the vertical temperature distribution in a semiconfining layer of an aquifer, in order to investigate the effect of horizontal water and heat flow. Their solution is an extension of the one-dimensional approach of Bredehoeft and Papadopulos (1965).

Incropera et al. (2007) describe analytical solutions for a series of technical problems, like heat exchanger systems.

Yang and Yeh (2008) formulated an analytical model for the radial heat transfer during the injection of hot water into a confined aquifer. Heat fluxes in the underlying and overlying rock are restricted to vertical conductive flux. Effects of heat dispersion are neglected.

Woods and Ortega (2011) formulated an analytical model to investigate the thermal response of a line of standing column wells and compared simulation results with results from numerical simulations.

Furthermore, for **borehole heat exchanger** (BHE) systems, various analytical models have been developed in the past. Based on analytical solutions of Eskilson and Claesson (1988) for BHEs and analytical expressions of Hellström (1991) for the thermal resistance of BHEs (Section 2.1.2.6), the **software EED** (Earth Energy Designer, current Version 3.16, BLOCON 2008) for the design of BHEs was developed. The software allows either the calculation of mean fluid temperature in BHEs, which are embedded in a medium with given properties (thermal conductivity, thermal capacity, mean ground surface temperature, geothermal heat flux) for given thermal load and BHE layout (diameter and length of borehole, type of BHE configuration), or the calculation of the required borehole length for given minimum and maximum temperatures of the fluid within the BHE. Further alternative software tools are **GLHEPRO** (current version 4.0, 2007) or **EWS** (Huber 2008, current version 4.0). Based on the cylinder source theory, Nagano et al. (2006) developed a **design and performance prediction tool** for ground-source heat pump systems. Lamarche and Beauchamp (2007) presented an analytical solution for the short-term analysis of BHEs with concentric cylindrical tubes. Based on numerical simulations, they demonstrated that the solution is also a good approximation for the U-tube configuration.

In the following, an **overview of analytical solutions**, relevant for the assessment of the thermal use of shallow groundwater systems, is given. We start with closed systems without local water withdrawal and reinjection, which, contrary to open systems, do not modify the original flow field. Thermal sources of the analytical solutions can be represented by a point source, an infinite line source (ILS), or a finite line source (FLS). These are shown schematically in Figure 3.1a through c. The temperature T is the sum of the ambient temperature without thermal use T_0 and a decrease (or increase) ΔT. The thermal velocity and the thermal diffusion/dispersion coefficient are indicated by an index t.

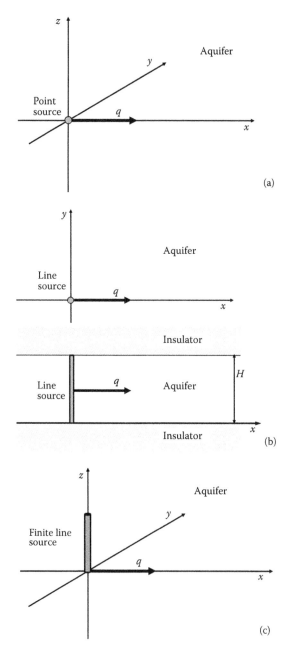

Figure 3.1 Thermal sources. (a) Point source in infinite aquifer with flow field q; (b) line source in aquifer layer bound by insulating layers with flow field q; (c) FLS in infinite aquifer with flow field q (schematic).

3.1 CLOSED SYSTEMS

3.1.1 Instantaneous point source— three-dimensional conduction

The three-dimensional **differential equation for heat conduction** with constant coefficients, and without internal sources or sinks, is

$$\frac{\partial T}{\partial t} = D_t \nabla^2 T = D_t \cdot \left(\frac{\partial^2 T}{\partial x^2} + \frac{\partial^2 T}{\partial y^2} + \frac{\partial^2 T}{\partial z^2} \right) \tag{3.1}$$

where D_t is the thermal diffusion coefficient with $D_t = \lambda_m / C_m$. The initial condition is $T(x, y, z, t = 0) = T_0$. A **quantity of energy** ΔE (J) injected instantaneously at the point (x_0, y_0, z_0) within an infinite three-dimensional aquitard produces a temperature distribution $T(x, y, z, t)$ given by an **instantaneous point source solution** (Carslaw and Jaeger 1959):

$$T(x,y,z,t) = T_0 + \frac{\Delta E}{8C_m (\pi D_t t)^{3/2}} \exp \left[-\frac{(x - x_0)^2 + (y - y_0)^2 + (z - z_0)^2}{4D_t t} \right] \tag{3.2}$$

In each coordinate direction x, y, and z, a bell-shaped temperature distribution is obtained. Accordingly, a negative injection corresponds to heat extraction. The point source may correspond to a small portion of a heat exchanger. Note that the initial temperature at the source location is infinite. This is due to the idealized condition of finite energy in a point.

3.1.2 Moving point source—three-dimensional conduction and advection

The three-dimensional **differential equation for heat conduction and advection** with constant coefficients, without internal sources or sinks, and for uniform groundwater flow in x-direction is

$$\frac{\partial T}{\partial t} = D_{t,L} \frac{\partial^2 T}{\partial x^2} + D_{t,T} \frac{\partial^2 T}{\partial y^2} + D_{t,T} \frac{\partial^2 T}{\partial z^2} - u_{t,x} \frac{\partial T}{\partial x} \tag{3.3}$$

The initial condition is $T(x, y, z, t = 0) = T_0$. The coefficient $D_{t,L}$ (m² s⁻¹) includes both thermal diffusion and dispersion. The **moving point source** with source strength $J = dE/dt$ (W) at a point located at (x_0, y_0, z_0) within an infinite three-dimensional aquifer corresponds to the problem of a **point**

source moving in the x-direction in an aquifer with zero flow (Carslaw and Jaeger 1959) with the temperature

$$T(x,y,z,t) = T_0 + \frac{J}{8C_m \left(\pi^3 D_{t,L} D_{t,T}^2\right)^{1/2}} \times \int_0^t \exp\left[-\left(\frac{(x - x_0 - u_t \cdot (t - t'))^2}{D_{t,L}}\right.\right.$$
$$\left.\left.+ \frac{(y - y_0)^2 + (z - z_0)^2}{D_{t,T}}\right)\frac{1}{4 \cdot (t - t')}\right] \cdot \frac{1}{(t - t')^{3/2}} dt'$$

(3.4)

The evaluation of the integral can be performed numerically. The temperature at the source location is infinite. The continuous point source may correspond to a small portion of a heat exchanger with continuous heat injection/extraction. In this case, the temperature has to be taken at the borehole wall, thus representing an approximation to the real situation.

3.1.3 ILS—two-dimensional conduction

Assuming the BHE as a vertical line source at location (x_0, y_0) with infinite length along the vertical (z_0) direction, we integrate Equation 3.2 along an infinite line, $-\infty < z_0 < \infty$, in order to get the **instantaneous line source** (Carslaw and Jaeger 1959):

$$T(x,y,z,t) = T_0 + \frac{\Delta E/H}{8C_m(\pi D_t t)^{3/2}} \int_{-\infty}^{\infty} \exp\left(-\frac{(x - x_0)^2 + (y - y_0)^2 + (z - z_0)^2}{4D_t t}\right) dz_0$$

(3.5)

where $\Delta E/H$ (J m^{-1}) is the heat energy per unit length of the borehole of length H (m), which is extended to the whole infinite length of the borehole of the model. Solving the previous integral results in

$$T(x,y,t) = T_0 + \frac{\Delta E/H}{4\pi\lambda_m t} \exp\left(-\frac{(x - x_0)^2 + (y - y_0)^2}{4D_t t}\right)$$

(3.6)

Integrating Equation 3.6 from $0 < t' < t$, we get the **continuous line source**

$$T(x,y,t) = T_0 + \frac{q_{tb}}{4\pi\lambda_m} \int_0^t \exp\left(-\frac{r^2}{4D_t(t - t')}\right) \frac{dt'}{(t - t')}$$

(3.7)

where $q_{tb} = J/H = (dE/dt)/H$ (W m^{-1}) is the **heat flow rate per unit length of the borehole** and $r^2 = (x - x_0)^2 + (y - y_0)^2$ is the radial coordinate. Introducing the dimensionless variable $u = r^2/4D_t(t - t')$ and the term $dt'/(t - t') = du/u$ results in

$$T(x,y,t) = T_0 + \frac{q_{tb}}{4\pi\lambda_m} \int_{r^2/(4D_t t)}^{\infty} \exp(-u)\frac{du}{u} \tag{3.8}$$

Moreover, making use of the definition of the **exponential integral**

$$-\text{Ei}(-x) = \int_x^{\infty} e^{-u}\frac{du}{u} \tag{3.9}$$

the solution for the **ILS** can be expressed as follows:

$$T(x,y,t) = T_0 - \frac{q_{tb}}{4\pi\lambda_m}\text{Ei}\left(-\frac{r^2}{4D_t t}\right) \tag{3.10}$$

In hydrological literature, the function $-\text{Ei}(-x)$ is also known as the well function $W(x)$. Equation 3.10 is, in particular, applicable for the evaluation of short-term geothermal field experiments such as thermal response tests, which usually last from 12 to 60 h (Signorelli et al. 2007). Introducing the **dimensionless temperature rise** $\Theta = 4\pi\lambda_m\Delta T/(J/H)$, where $\Delta T = T - T_0$, the **dimensionless radial coordinate** $R = r/L$, where L is the length scale of interest, and the **Fourier number** Fo, which can be interpreted as dimensionless time, with

$$\text{Fo} = D_t t/L^2 \tag{3.11}$$

we can express Equation 3.10 in dimensionless form as follows:

$$\Theta_{ILS} = -\text{Ei}\left[-\frac{R^2}{4\text{Fo}}\right] \tag{3.12}$$

The instantaneous and the continuous ILS models can be directly applied for a **thin BHE** sufficiently far away from the upper and lower ends of the BHE. Moreover, it can be applied for a **finite soil layer** of thickness H (length of the BHE), which is **limited by a thermally insulating top and bottom layer** (Figure 3.1b).

Based on the infinite line-source model and the thermal load of a BHE system, Michopoulos and Kyriakis (2009) developed and evaluated a model to predict the temperature at the exit of a vertical ground heat exchanger.

The **coded function** (MATLAB script) of the **ILS** model (Equation 3.10) is listed as *T_ILS.m*. It can be found at http://www.crcpress.com/product/isbn/9781466560192. A program to visualize the two-dimensional temperature field response for a single borehole with a continuous heat flow rate is available as *closedsys.m*. As an example, Figure 3.2 shows the radial temperature propagation after 90 days of operating a BHE with specific heat extraction $q_{tb} = J/H$ of 50 W m^{-1}.

Solutions to the ILS model for the **time-dependent heat input function of a group of BHEs** can be obtained by applying the superposition principle over all BHEs and over a series of time increments. Figure 3.3 shows the seasonal heat input defined by a cosine function as

$$J_{BHE}(x = 0, t) = J_{ampl}\cos(\alpha - \omega t) \tag{3.13}$$

for $\alpha = 0$ (phase shift) and the heat input amplitude $J_{ampl} = 62.8$ W m^{-1}. The symbol ω (s^{-1}) is the angular frequency, where $\omega = 2\pi/\tau$ with the length of the period τ.

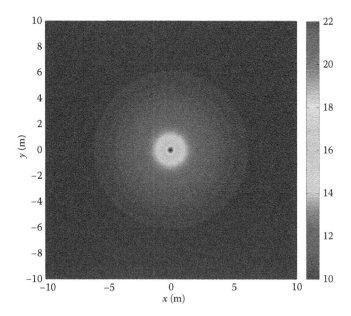

Figure 3.2 **(See color insert.)** Temperature field for a single BHE with constant energy extraction after 90 days. ILS model ($q_{tb} = J/H = 50$ W m^{-1}, $T_0 = 10°$C, $D_t = 9 \times 10^{-7}$ m^2 s^{-1}).

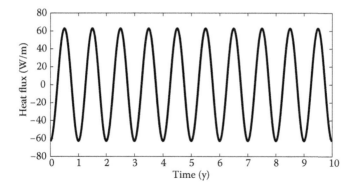

Figure 3.3 Seasonal cosine heat input function (example).

The temperature field of **interacting boreholes** is calculated by summing up the temperature response of individual BHEs:

$$\Delta T = \sum_{i=1}^{N} \Delta T_i \qquad (3.14)$$

where N denotes the number of BHEs. A MATLAB program to visualize a two-dimensional temperature field response for multiple boreholes is listed as *closedsys_mBHE.m*. Figure 3.4a through d shows the seasonal heat input and temperature maps for the times 10.0, 10.25, 10.5, and 10.75 years after the operation began. The system has almost reached quasi-steady state. The geometric arrangement and operation mode adjustment in low enthalpy geothermal fields for heating was studied by Beck et al. (2013) using similar models.

3.1.4 Infinite cylindrical source— two-dimensional conduction

For cases in which the **radius of the BHE** (r_0) **is important, the source is considered as a cylindrical surface** and the heat flow rate is applied at $r = r_0$. Ingersoll et al. (1954) presented the following equation:

$$T(x,y,t) = T_0 + \frac{q_{tb}}{\pi^2 \lambda_m} \int_0^\infty \frac{e^{-\beta^2 Fo} - 1}{J_1^2(\beta) + Y_1^2(\beta)} \times [J_0(R\beta)Y_1(\beta) - J_1(\beta)Y_0(R\beta)] \frac{d\beta}{\beta^2}$$

$$(3.15)$$

where $R = r/r_0$ $(L = r_0)$ is the dimensionless cylindrical radius. The functions J_0 and J_1 are Bessel functions of the first kind and of orders zero and one, whereas Y_0 and Y_1 are Bessel functions of the second kind of orders zero and one. Equation 3.15 is difficult to evaluate. A simpler expression can be

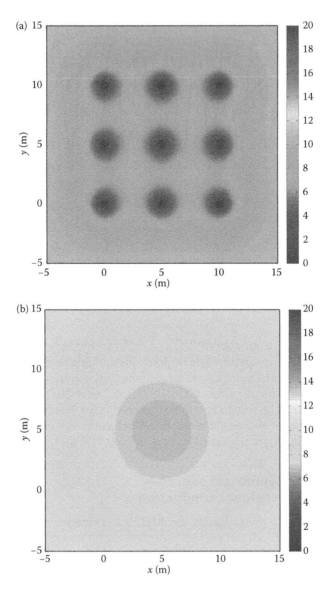

Figure 3.4 **(See color insert.)** Heat exchanger group 3 × 3 calculated with ILS model with seasonal cosine heat input. (a) Map after 10.0 years; (b) 10.25 years; (c) 10.5 years; and (d) 10.75 years.

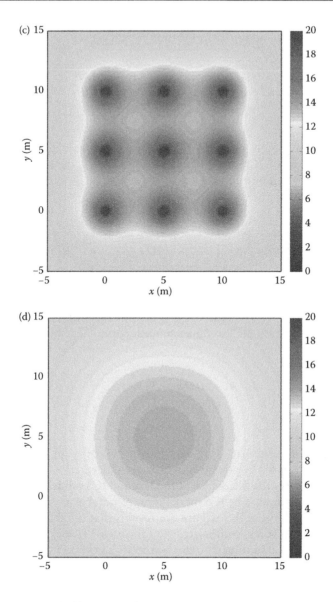

Figure 3.4 (Continued) **(See color insert.)** Heat exchanger group 3 × 3 calculated with ILS model with seasonal cosine heat input. (a) Map after 10.0 years; (b) 10.25 years; (c) 10.5 years; and (d) 10.75 years.

derived by expressing the line source model (Equation 3.10) in radial coordinates (r, φ_r) for a line source at location r_0 and φ_r':

$$T(r,\varphi,t) = T_0 - \frac{q_{tb}}{4\pi\lambda_m} \text{Ei}\left(-\frac{r^2 + r_0^2 - 2rr_0 \cos(\varphi_r - \varphi_r')}{4D_t t}\right) \tag{3.16}$$

where r and φ_r denote the radial and angular coordinates, respectively. Integrating Equation 3.16 around a circle of radius r_0, the **infinite cylindrical source** (ICS) can be expressed as follows (Man et al. 2010):

$$T(r,t) = T_0 - \frac{q_{tb}}{4\pi\lambda_m} \int_0^\pi \frac{1}{\pi} \text{Ei}\left(-\frac{r^2 + r_0^2 - 2rr_0 \cos\varphi_r'}{4D_t t}\right) d\varphi_r' \tag{3.17}$$

The integral can be evaluated numerically. Introducing the dimensionless time $\text{Fo} = D_t t / r_0^2$ $(L = r_0)$, we can express the **dimensionless temperature rise** according to Equation 3.17 in dimensionless form:

$$\Theta_{\text{ICS}} = -\frac{1}{\pi} \int_0^\pi \text{Ei}\left(-\frac{R^2 + 1 - 2R\cos\varphi_r'}{4\,\text{Fo}}\right) d\varphi_r' \tag{3.18}$$

Equations 3.17 and 3.18 were first introduced by Man et al. (2010) to simulate heat transfer by pile ground heat exchangers.

Figure 3.5 shows the **difference of the ILS and ICS models**, especially for short time simulations. The ICS model is more suitable for short time simulations compared to the ILS. Figure 3.6 reveals that the effect of assuming the borehole as an ILS becomes irrelevant for $\text{Fo} \geq 8$ $(L = r_0)$ when $R = 1$ (at the borehole wall) assuming $\Theta_{\text{ILS}}/\Theta_{\text{ICS}} > 0.99$ as a criterion. Eskilson (1987) states that the ILS model is valid for $\text{Fo} > 5$. In comparison, Ingersoll et al. (1954) were more strict and stated that the ILS model is only valid for $\text{Fo} > 20$. Philippe et al. (2009) investigated the validity range of analytical solutions to the ILS, FLS, and ICS models. Bernier et al. (2004) suggested a technique to aggregate heating and cooling loads when using the cylindrical source models to perform annual hourly energy simulations of ground coupled heat pump systems.

The **coded function** (MATLAB script) of the **ICS** model (Equation 3.17) is available as *T_ICS.m*. As an example, a BHE with specific heat extraction $q_{tb} = J/H$ of 50 W m^{-1} and a radius of 0.1 m in an aquifer with a thermal

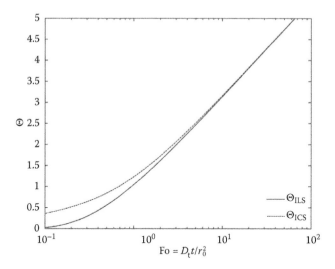

Figure 3.5 Dimensionless temperature as function of dimensionless time Fo ($R = 1.0$, $L = r_0$).

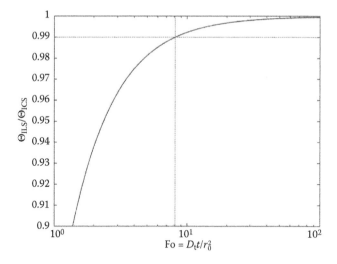

Figure 3.6 Ratio of ILS and ICS models over the dimensionless time Fo ($R = 1.0$, $L = r_0$).

diffusion coefficient of 9.0×10^{-7} m^2 s^{-1} results in a temperature change at the borehole wall of -1.6 K after 3 h (Fo = 1.0) when using the ILS model. In comparison, the ICS yields a temperature change of -1.9 K. Due to this discrepancy, the use of the more appropriate ICS is favorable for these specific conditions.

3.1.5 FLS—three-dimensional conduction

In order to account for axial effects, the borehole must be considered to have finite length. Integrating Equation 3.2 between $0 \le t' \le t$, we get the **continuous point source** (Carslaw and Jaeger 1959):

$$T(r,t) = T_0 + \frac{J}{8C_m \cdot (\pi D_t)^{3/2}} \int_0^t \exp\left(\frac{r'^2}{4D_t \cdot (t-t')}\right) \frac{dt'}{(t-t')^{3/2}} \tag{3.19}$$

where $J = dE/dt$ (W) is the strength of the continuous point source and

$$r' = \sqrt{(x-x_0)^2 + (y-y_0)^2 + (z-z_0)^2} = \sqrt{r^2 + (z-z_0)^2} \tag{3.20}$$

is the radial coordinate. Applying the change of variables $\tau = (t - t')^{-1/2}$ and $dt'/(t - t')^{-3/2} = 2d\tau$ results in

$$T(r,t) = T_0 + \frac{J}{4C_m(\pi D_t)^{3/2}} \int_{1/\sqrt{t}}^\infty \exp\left(-\frac{r'^2\tau^2}{4D_t}\right) d\tau \tag{3.21}$$

and finally in

$$T(x,y,z,t) = T_0 + \frac{J}{4\pi\lambda_m r'} \operatorname{erfc}\left(\frac{r'}{\sqrt{4D_t t}}\right) \tag{3.22}$$

where erfc(x) is the **complementary error function**:

$$\operatorname{erfc}(x) = \frac{2}{\sqrt{\pi}} \int_x^\infty \exp(-y^2) \, dy \tag{3.23}$$

When t approaches infinity, Equation 3.22 can be approximated by the **steady-state point source solution**:

$$T(x,y,z) = T_0 + \frac{J}{4\pi\lambda_m r'} \tag{3.24}$$

The contributions of point sources of equal energy injection/extraction making up a line source can be added (Eskilson 1987; Lamarche and Beauchamp 2007; Marcotte et al. 2010; Zeng et al. 2002), and the constant surface temperature boundary condition can be satisfied by applying the method of images (Figure 3.7). Applying this method (Eskilson 1987; Zeng et al. 2002) to Equation 3.22 yields the FLS model:

$$T(x,y,z,t)=T_0+\frac{q_{tb}}{4\pi\lambda_m}\left\{\int_0^H \frac{\text{erfc}\left(r'/\sqrt{4D_t t}\right)}{r'}dz_0 - \int_{-H}^0 \frac{\text{erfc}\left(r'/\sqrt{4D_t t}\right)}{r'}dz_0\right\}$$

(3.25)

where H is the borehole length.

For **steady-state conditions**, Equation 3.25 reduces to

$$T(x,y,z)=T_0+\frac{q_{tb}}{4\pi\lambda_m}\ln\left(\frac{H-z+\sqrt{r^2+(H-z)^2}}{H+z+\sqrt{r^2+(H+z)^2}}\cdot\frac{2z^2+2z\sqrt{r^2+z^2}+r^2}{r^2}\right)$$

(3.26)

Introducing the dimensionless time Fo $= D_t t/H^2$ ($L = H$) and the dimensionless coordinates $R = r/H$, $Z = z/H$, and $Z' = z_0/H$, we can express the transient FLS model (Equation 3.25) **in dimensionless form**:

$$\Theta_{FLS}=\int_0^1 \frac{\text{erfc}\left[\sqrt{R^2+(Z-Z')^2}/2\sqrt{Fo}\right]}{\sqrt{R^2+(Z-Z')^2}}dZ'$$
$$-\int_{-1}^0 \frac{\text{erfc}\left[\sqrt{R^2+(Z-Z')^2}/2\sqrt{Fo}\right]}{\sqrt{R^2+(Z-Z')^2}}dZ'$$

(3.27)

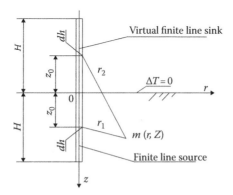

Figure 3.7 Representation of the FLS. (Modified after Zeng, H.Y. et al. *Heat Transfer-Asian Research* 31 (7), 558–567, 2002.)

and also the steady-state Equation 3.26:

$$\Theta_{FLS\_steady-state} = \ln\left[\frac{1-Z+\sqrt{R^2+(1-Z)^2}}{1+Z+\sqrt{R^2+(1+Z)^2}}\cdot\frac{2Z^2+2Z\sqrt{R^2+Z^2}+R^2}{R^2}\right] \quad (3.28)$$

From Figure 3.8, it can be seen that **axial effects become important for long time simulations**. The shorter the borehole length, the higher the discrepancy between the ILS and FLS models. Figure 3.9 shows that the axial effects are negligible for Fo < 0.052 ($L = H$) when $R = 0.005$ ($H = 10$ m, $r = r_0 = 0.05$ m) assuming $\Theta_{FLS}/\Theta_{ILS} > 0.9$ as a criterion. For $R = 0.0005$ ($H = 100$ m, $r = r_0 = 0.05$ m), axial effects are negligible for Fo < 0.065. Eskilson (1987) is more restrictive and states that the ILS model is valid for Fo < 0.01.

Bandos et al. (2009) present a solution to the FLS model, which takes into account the prevailing geothermal gradient and arbitrary ground surface temperatures. Marcotte et al. (2010) investigated the importance of axial effects by comparing solutions of the ILS and FLS models. Cui et al. (2006) formulated an inclined FLS analytical solution.

The **coded functions** (MATLAB scripts) of the **FLS** model for transient (Equation 3.25) and steady-state conditions (Equation 3.26) are listed as

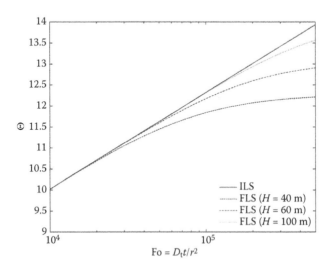

Figure 3.8 Dimensionless temperature response at the borehole wall over dimensionless time for different borehole lengths H (m) (R = 0.005, L = H = 10 m, r = r₀ = 0.05 m, z = 0.5 × H).

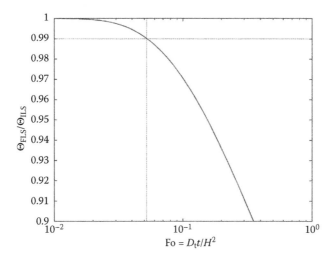

Figure 3.9 Ratio of the FLS and the ILS models over the dimensionless time Fo (R = 0.005, L = H = 10 m, r = 0.05 m, z = 0.5 × H).

T_FLS.m and *T_FLSs.m*, respectively. A program to visualize the temperature at the borehole wall over time for a single borehole with a continuous heat flow rate is listed as *closedsys_Tb.m*. As an example, Figure 3.10 shows the temperature response at the borehole wall for a BHE with specific heat extraction $q_{tb} = J/H$ of 50 W m^{-1} using the ILS and FLS models.

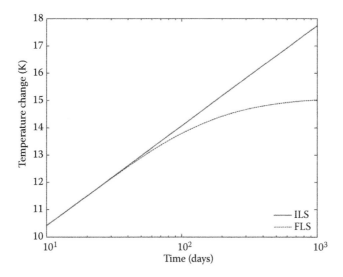

Figure 3.10 Temperature response at the borehole wall over time (H = 10 m, z = 0.5 × H, D_t = 9 × 10^{-7} m^2 s^{-1}).

3.1.6 Finite cylindrical source— three-dimensional conduction

First, we express the **continuous point source model** (Equation 3.22) in radial coordinates as follows:

$$T(r, \varphi_{r3}, z, t) = T_0 + \frac{J}{4\pi\lambda_m} \frac{\mathrm{erfc}\left(\sqrt{r^2 + r_0^2 - 2rr_0\cos(\varphi_r - \varphi_r') + (z - z_0)^2}\Big/\sqrt{4D_t t}\right)}{\sqrt{r^2 + r_0^2 - 2rr_0\cos(\varphi_r - \varphi_r') + (z - z_0)^2}}$$

(3.29)

Then we integrate Equation 3.29 around a circle of radius r_0 in order to get the **continuous ring source** model:

$$T(r, z, t) = T_0 + \frac{J}{4\pi\lambda_m} \int_0^\pi \frac{\mathrm{erfc}\left[\sqrt{r^2 + r_0^2 - 2rr_0\cos\varphi_r' + (z - z_0)^2}\Big/\sqrt{4D_t t}\right]}{\pi\sqrt{r^2 + r_0^2 - 2rr_0\cos\varphi_r' + (z - z_0)^2}} d\varphi_r'$$

(3.30)

For **steady-state conditions**, Equation 3.30 reduces to

$$T(r, z) = T_0 + \frac{J}{4\pi\lambda_m} \int_0^\pi \frac{1}{\pi\sqrt{r^2 + r_0^2 - 2rr_0\cos\varphi_r' + (z - z_0)^2}} \varphi_r'$$

(3.31)

Integrating over the borehole length and adding the upper constant temperature boundary condition by applying the method of images to Equations 3.30 and 3.31 yields the **finite cylindrical source (FCS)** model:

$$T(r, z, t) = T_0 + \frac{q_{tb}}{4\pi\lambda_m} \left\{ \int_0^H \int_0^\pi \frac{\mathrm{erfc}\left[r'/\sqrt{4D_t t}\right]}{\pi r'} d\varphi_r' \, dz' - \int_{-H}^0 \int_0^\pi \frac{\mathrm{erfc}\left[r'/\sqrt{4D_t t}\right]}{\pi r'} d\varphi_r' \, dz' \right\}$$

(3.32)

$$T(r, z) = T_0 + \frac{q_{tb}}{4\pi\lambda_m} \left\{ \int_0^H \int_0^\pi \frac{1}{\pi r'} d\varphi_r' \, dz' - \int_{-H}^0 \int_0^\pi \frac{1}{\pi r'} d\varphi_r' \, dz' \right\}$$

(3.33)

where $r' = \sqrt{r^2 + r_0^2 - 2rr_0 \cos \varphi_r' + (z - z')^2}$. Similar equations have been presented by Man et al. (2010). Expressing Equations 3.32 and 3.33 in dimensionless form yields

$$\Theta_{FCS} = \int_0^1 \int_0^\pi \frac{\text{erfc}\left[R'/2\sqrt{Fo}\right]}{\pi R'} d\varphi_r' \, dZ' - \int_{-1}^0 \int_0^\pi \frac{\text{erfc}\left[R'/2\sqrt{Fo}\right]}{\pi R'} d\varphi_r' \, dZ' \quad (3.34)$$

$$\Theta_{FCS\_steady-state} = \int_0^1 \int_0^\pi \frac{1}{\pi R'} d\varphi_r' \, dz' - \int_{-1}^0 \int_0^\pi \frac{1}{\pi R'} d\varphi_r' \, dz' \quad (3.35)$$

where $R' = \sqrt{R^2 + R_0^2 - 2RR_0 \cos \varphi_r' + (Z - Z')^2}$ and $R_0 = r_0/L$.

The behavior with respect to the influence of the borehole radius in a conduction-dominated problem is similar to the one shown in Figures 3.5 and 3.6.

The **coded functions** (MATLAB script) of the **FCS** model for transient (Equation 3.32) and steady-state (Equation 3.33) conditions are listed as *T_FCS.m* and *T_FCSs.m*, respectively.

3.1.7 Moving ILS—two-dimensional conduction and advection

An ILS in an aquifer with uniform flow according to Figure 3.1b corresponds to the moving ILS (MILS). Applying the moving source theory to Equation 3.7 yields the analytical solution for the response of a constant line source of infinite length along the vertical direction with a continuous heat flow rate $q_{tb} = J/H$ per unit length of the borehole, or the **MILS model**:

$$T(x,y,t) = T_0 + \frac{q_{tb}}{4\pi\lambda_m} \int_0^t \exp\left(-\frac{\{(x - u_t \cdot (t - t'))\}^2 + y^2}{4D_t \cdot (t - t')}\right) \frac{dt'}{(t - t')} \quad (3.36)$$

For the sake of simplicity, the source is located at $x_0 = y_0 = 0$. Applying the following change of variable $\psi = r^2/4D_t \cdot (t - t')$, $dt'/(t - t') = d\psi/\psi$, and $r = \sqrt{x^2 + y^2}$ yields

$$T(x,y,t) = T_0 + \frac{q_{tb}}{4\pi\lambda_m} \exp\left[\frac{u_t x}{2D_t}\right] \int_{r^2/4D_t t}^\infty \exp\left[-\psi - \frac{u_t^2 r^2}{16D_t^2 \psi}\right] \frac{d\psi}{\psi} \quad (3.37)$$

For **steady state-conditions,** Equation 3.37 becomes (Carslaw and Jaeger 1959)

$$T(x,y) = T_0 + \frac{q_{tb}}{2\pi\lambda_m} \exp\left[\frac{u_t x}{2D_t}\right] K_0\left[\frac{u_t\sqrt{x^2+y^2}}{2D_t}\right]$$
(3.38)

in which K_0 is the modified Bessel function of the second kind of order zero. Equations 3.37 and 3.38 have previously been used by Sutton et al. (2003), Zubair and Chaudhry (1996), and Diao et al. (2004) to calculate the ground resistance, temperature distributions for time-dependent energy extraction/injection, and the effects of groundwater advection on ground-source heat pump systems.

Introducing the thermal Peclet number $Pe = u_t L/D_t$, we can express the MILS model (Equation 3.37) **in dimensionless form** as follows:

$$\Theta_{MILS} = \exp\left[\frac{Pe}{2} R\cos\varphi_r\right] \int_{R^2/4Fo}^{\infty} \exp\left[-\psi - \frac{Pe^2 R^2}{16\psi}\right] \frac{d\psi}{\psi}$$
(3.39)

$$\Theta_{MILS\_steady\text{-}state} = 2\exp\left[\frac{Pe}{2} R\cos\varphi_r\right] K_0\left[\frac{Pe}{2} R\right]$$
(3.40)

Recall that φ_r is the angular coordinate (polar angle) and $R = r/L$. If groundwater flow is present, temperature distribution in the x–y plane is not symmetrical with respect to the polar angle. Figure 3.11 shows the dimensionless temperature distribution using Equation 3.40.

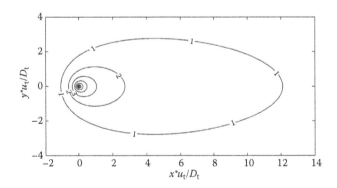

Figure 3.11 Steady-state dimensionless isotherms considering background groundwater flow (q = 1.0 × 10⁻⁶ m s⁻¹, D_t = 9 × 10⁻⁷ m² s⁻¹).

The **coded functions** (MATLAB script) of the **MILS** model for transient (Equation 3.37) and steady-state (Equation 3.38) conditions are listed as *T_MILS.m* and *T_MILSs.m*, respectively. A program to visualize the two-dimensional temperature field response for a single borehole with continuous heat flow rate in an aquifer with uniform horizontal groundwater flow is listed as *closedsys.m*. As an example, Figure 3.12 shows the two-dimensional temperature response for steady-state conditions of a BHE with specific heat extraction $q_{tb} = J/H$ of 50 W m^{-1}. The aquifer has an initial temperature $T_0 = 10°C$, a thermal diffusion coefficient of 9×10^{-7} m^2 s^{-1}, and a uniform groundwater flow velocity of 1.0×10^{-6} m s^{-1}. An example of a two-dimensional temperature field of multiple BHEs in an aquifer with background groundwater flow is shown in Figure 3.13. The **program** is listed as *closedsys_mBHE.m*.

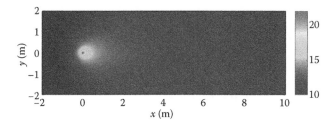

Figure 3.12 **(See color insert.)** Temperature field for a single BHE with constant energy extraction after 90 days ($q_{tb} = J/H = 50$ W m^{-1}, $T_0 = 10°C$, $q = 1.0 \times 10^{-6}$ m s^{-1}, $D_t = 9 \times 10^{-7}$ m^2 s^{-1}).

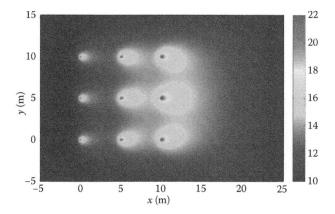

Figure 3.13 **(See color insert.)** Temperature field of multiple interacting BHEs with constant energy extraction after 90 days ($q_{tb} = J/H = 50$ W m^{-1}, $T_0 = 10°C$, $q = 1.0 \times 10^{-6}$ m s^{-1}, $D_t = 9 \times 10^{-7}$ m^2 s^{-1}).

In order to compute the mean temperature around a borehole in an aquifer with uniform horizontal groundwater flow, the integral average of the temperature response of a circle of radius r_0 must be estimated (Diao et al. 2004).

Taking into account the following definition of the modified Bessel function of the first kind and of order zero

$$I_0(u) = \frac{1}{\pi} \int_0^\pi \exp(u \cos \varphi') d\varphi' \tag{3.41}$$

the mean temperature at the borehole wall ($r = r_0$) for the MILS for transient (Equation 3.37) and steady-state conditions (Equation 3.38) is as follows:

$$T(r_0,t) = T_0 + \frac{q_{tb}}{4\pi\lambda_m} I_0 \left[\frac{u_t r}{2D_t} \right] \int_{r^2/4D_t t}^\infty \exp\left[-\psi - \frac{u_t^2 r^2}{16 D_t^2 \psi} \right] \frac{d\psi}{\psi} \tag{3.42}$$

$$T(r_0) = T_0 + \frac{q_{tb}}{2\pi\lambda_m} I_0 \left[\frac{u_t r}{2D_t} \right] K_0 \left[\frac{u_t r}{2D_t} \right] \tag{3.43}$$

In dimensionless form, we get

$$\Theta_{MILS} = I_0 \left[\frac{Pe}{2} R \right] \int_{R^2/4Fo}^\infty \exp\left[-\psi - \frac{Pe^2 R^2}{16\psi} \right] \frac{d\psi}{\psi} \tag{3.44}$$

$$\Theta_{MILS\_steady\text{-}state} = 2 I_0 \left[\frac{Pe}{2} R_0 \right] K_0 \left[\frac{Pe}{2} R_0 \right] \tag{3.45}$$

Although a BHE consists of a buried pipe, which commonly is embedded in grouting material, the approximation by a line source is commonly accepted as an approximation in heat transport models of ground-source systems (Diao et al. 2004; Eskilson 1987; Sutton et al. 2003).

The **coded functions** (MATLAB script) of the **MILS** model for computing the mean temperature at the borehole wall (Equations 3.42 and 3.43) are *T_MILSc.m* and *T_MILScs.m*, respectively. An example of the

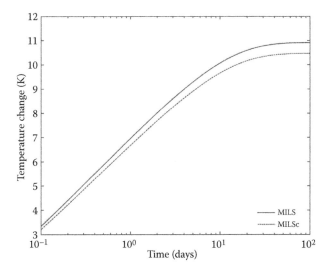

Figure 3.14 Temperature response at the borehole wall over time ($q_{tb} = J/H = 50$ W m^{-1}, $q = 1.0 \times 10^{-6}$ m s^{-1}, $D_t = 9 \times 10^{-7}$ m^2 s^{-1}). MILS: moving ILS ($x = r_0$); MILSc: MILS with mean temperature at the borehole wall ($r = r_0$).

temperature response at the borehole wall is shown in Figure 3.14 using Equations 3.37 and 3.42 with constant energy extraction. Note that in Figure 3.14, Equation 3.42 computes the average temperature around the borehole wall, whereas Equation 3.37 computes the temperature at $x = r_0$ and $y = 0$.

In order to **consider thermal dispersion**, we express the **instantaneous line source** equation (Equation 3.6) for **anisotropic material** (Carslaw and Jaeger 1959) and apply the moving source theory, which yields

$$T(x,y,t) = T_0 + \frac{\Delta E/H}{4\pi C_m \sqrt{D_{t,L} D_{t,T}}\, t} \exp\left[-\frac{[x - u_t t]^2}{4 D_{t,L} t} - \frac{y^2}{4 D_{t,T} t} \right] \qquad (3.46)$$

where $D_{t,L}$ and $D_{t,T}$ are the longitudinal and transversal thermal diffusivity coefficients (Equation 2.93), respectively, which include thermal dispersion effects, given by

$$D_{t,L} = D_t + \beta_L u_t \qquad (3.47)$$

$$D_{t,T} = D_t + \beta_T u_t \qquad (3.48)$$

Let us now consider a continuous line source, where a constant heat flow rate $q_{tb} = J/H$ is continuously injected/extracted. Integrating Equation 3.46 over the time interval $(0, t)$:

$$T(x,y,t) = T_0 + \frac{q_{tb}}{4\pi C_m \sqrt{D_{t,L} D_{t,T}}} \int_0^t \exp\left[\frac{-[x - u_t \cdot (t-t')]^2}{4D_{t,L} \cdot (t-t')} - \frac{y^2}{4D_{t,T} \cdot (t-t')}\right]$$

$$\times \frac{dt'}{(t-t')}$$

$$(3.49)$$

and applying the change of variable yields

$$T(x,y,t) = T_0 + \frac{q_{tb}}{4\pi C_m \sqrt{D_{t,L} D_{t,T}}} \exp\left[\frac{u_t x}{2D_{t,L}}\right]$$

$$\times \int\limits_{\left(\frac{x^2}{4D_{t,L}t} + \frac{y^2}{4D_{t,T}t}\right)}^{\infty} \exp\left[-\psi - \left(\frac{x^2}{D_{t,L}} + \frac{y^2}{D_{t,T}}\right)\frac{u_t^2}{16D_{t,L}\psi}\right]\frac{d\psi}{\psi} \quad (3.50)$$

Metzger et al. (2004) used this analytical solution to estimate thermal dispersion coefficients for a packed bed of glass spheres. Hecht-Méndez et al. (2013) applied superposition of Equation 3.50 to optimize multiple BHE operation in a BHE field

$$T(x,y) = T_0 + \frac{J/H}{2\pi C_m \sqrt{D_{t,L} D_{t,T}}} \exp\left[\frac{u_t x}{2D_{t,L}}\right] K_0\left[\frac{u_t}{2}\sqrt{\frac{D_{t,T}x^2 + D_{t,L}y^2}{D_{t,L}^2 D_{t,T}}}\right] \quad (3.51)$$

In Equations 3.49 and 3.50, thermal dispersivities are set to zero ($\beta_L = \beta_T = 0$) when thermal dispersion is ignored. This yields the analytical solutions given by Equations 3.37 and 3.38.

The **coded functions** (MATLAB scripts) of the **MILS** model for transient (Equation 3.35) and steady-state (Equation 3.36) conditions, considering thermal dispersion, are listed as *T_MILSd.m* and *T_MILSsd.m*, respectively.

As an example, Figure 3.15 shows the two-dimensional temperature field response of a BHE with specific heat extraction of 50 W m^{-1} for different values of thermal dispersivity. The length of a plume is defined via an isothermal contour ΔT, as the distance between the source and the intersection of this isothermal contour with the x-axis. Increase in dispersivity

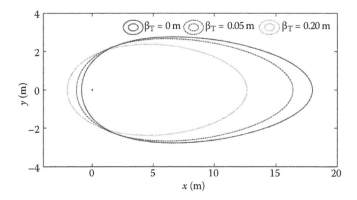

Figure 3.15 Steady-state temperature field for a single BHE with constant energy extraction for different thermal dispersivities ($\Delta T = 1$ K, $q_{tb} = J/H = 50$ W m^{-1}, $q = 1.0 \times 10^{-6}$ m s^{-1}, $D_t = 9 \times 10^{-7}$ m^2 s^{-1}). (Modified after Molina-Giraldo, N. et al. *International Journal of Thermal Sciences* 50 (7), 1223–1231, 2011.)

yields shorter temperature plumes for the given ΔT. The relative sensitivity of the temperature change near a BHE to longitudinal dispersivity almost disappears for long-term simulation. In contrast, according to Equation 3.51, sensitivity to transverse dispersivity grows with simulated time.

For steady-state conditions, an approximation can be made in order to calculate the **length of the temperature plume** by using an approximation of the modified Bessel function of the second kind of order zero, $K_0(u)$ (Carslaw and Jaeger 1959):

$$u^{0.5} \exp(u) K_0(u) \approx \sqrt{\frac{\pi}{2}} \left(1 - \frac{1}{8u} \right) \tag{3.52}$$

where u is the argument of the Bessel function. Substituting Equation 3.52 into Equation 3.51 and solving for **the temperature plume length** (L_p) yields

$$L_p = \left(\frac{(J/H)^2}{8\pi (C_m)^2 D_{t,T} u_t \Delta T^2} \right) \left(1 \pm \sqrt{1 - \frac{8\pi \cdot (C_m)^2 D_{t,L} D_{t,T} \Delta T^2}{(J/H)^2}} \right) \tag{3.53}$$

where ΔT is the value of the isothermal contour. Equation 3.53 has been previously used by Molina-Giraldo et al. (2011a) to evaluate the effect of thermal dispersion in temperature plumes from vertical ground-source heat pump systems. Ham et al. (2004) used it to estimate the effect of dispersion on solute transport.

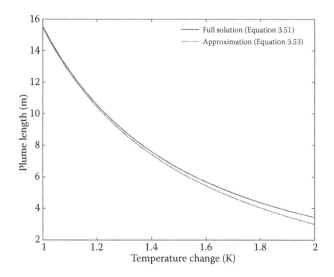

Figure 3.16 Plume length for steady-state conditions as a function of temperature change ($q = 1 \times 10^{-6}$ m s^{-1}, $D_t = 9 \times 10^{-7}$ m^2 s^{-1}, $\beta_L = 1.0$ m, $\beta_T = 0.1$ m, $q_{tb} = J/H = -50$ W m^{-1}, $y = 0$ m). (Modified after Molina-Giraldo, N. et al. *International Journal of Thermal Sciences* 50 (7), 1223–1231, 2011.)

The **coded function** (MATLAB script) of the **MILS** model to compute the length of the temperature plume at steady-state conditions (Equation 3.53) is available as *T_PL.m*.

Equation 3.53 can be employed to compute the length of a temperature plume (L_p) for a given isothermal contour (ΔT) under steady-state conditions. This is an estimate, which is valid only for $u \gg 1$. According to Abramowitz and Stegun (1964), the relative error of this approximation is around 0.01 when $u > 3$. Figure 3.16 shows an exemplary comparison of the full solution (Equation 3.50) with the approximation (Equation 3.53). The relative error for this specific example is about 8% for a $\Delta T = 2$ K.

3.1.8 Moving FLS—three-dimensional conduction and advection

The temperature response at a given time t due to an energy flux J extracted/injected by a **continuous point source** after applying the moving source theory to Equation 3.2 (Carslaw and Jaeger 1959) is

$$T(x,y,z,t) = T_0 + \frac{J}{8C_m \cdot (\pi D_t)^{3/2}} \int_0^t \exp\left[-\frac{\{(x - u_t \cdot (t - t'))\}^2 + y^2 + z^2}{4D_t \cdot (t - t')} \right] \frac{dt'}{(t - t')^{3/2}}$$

$$(3.54)$$

For the sake of simplicity, the source is located at $x_0 = y_0 = 0$. Applying the change of variable, $\psi = r'/2\sqrt{D_t \cdot (t - t')}$ and $dt'/(t - t')^{3/2} = 4\sqrt{D_t}/r' d\psi$, yields the moving point source equation for a continuous injection:

$$T(x,y,z,t) = T_0 + \frac{J}{2\pi^{3/2}\lambda_m r'} \exp\left[\frac{u_t x}{2D_t}\right] \int_{r'/\sqrt{4D_t t}}^{\infty} \exp\left[-\psi^2 - \frac{u_t^2 r'^2}{16 D_t^2 \psi^2}\right] d\psi$$

(3.55)

where $r' = \sqrt{x^2 + y^2 + (z - z_0)^2} = \sqrt{r^2 + (z - z_0)^2}$. For steady-state conditions, Equation 3.55 becomes

$$T(x,y,z,t) = T_0 + \frac{J}{4\pi\lambda_m r'} \exp\left[-\frac{u_t \cdot (r' - x)}{2D_t}\right]$$

(3.56)

Applying the following change of variable, $\phi = \psi^2$, to Equation 3.55 yields

$$T(x,y,z,t) = T_0 + \frac{J}{4\pi^{3/2}\lambda_m r'} \exp\left[\frac{u_t x}{2D_t}\right] \int_{r'^2/4D_t t}^{\infty} \exp\left[-\phi - \frac{u_t^2 r'^2}{16 D_t^2 \phi}\right] \frac{d\phi}{\sqrt{\phi}}$$

(3.57)

The integral of Equation 3.57 can be expressed as the generalized incomplete Gamma function (Chaudhry and Zubair 1994):

$$\Gamma(1/2, u_1; u_2) = \int_{u_1}^{\infty} \frac{1}{\sqrt{\phi}} \exp\left[-\phi - \frac{u_2}{\phi}\right] d\phi; \quad u_1 = \frac{r'^2}{4D_t t}; \quad u_2 = \frac{u_t^2 r'^2}{16 D_t^2}$$

(3.58)

From Equations 3.57 and 3.58, we have the following equation:

$$T(x,y,z,t) = T_0 + \frac{J}{4\pi^{3/2}\lambda_m} \exp\left[\frac{u_t x}{2D_t}\right] \frac{\Gamma(1/2, u_1, u_2)}{r'}$$

(3.59)

In order to account for axial effects and constant ground surface temperature conditions, the method of images (Carslaw and Jaeger 1959; Eskilson

1987) is applied to Equation 3.59, resulting in the **moving FLS** (MFLS) model:

$$T(x,y,z,t)=T_0+\frac{J/H}{4\pi\lambda_m}\exp\left[\frac{u_r x}{2D_t}\right]\left\{\int_0^H\frac{\Gamma(1/2,u_1;u_2)}{\sqrt{\pi}r'}\,dz_0-\int_{-H}^0\frac{\Gamma(1/2,u_1;u_2)}{\sqrt{\pi}r'}\,dz_0\right\}$$

(3.60)

The generalized incomplete Gamma function can be approximated by the following function (Chaudhry and Zubair 1994):

$$\Gamma(1/2,u_1;u_2)\cong\frac{1}{2}\sqrt{\pi}\cdot\left[\exp\left(-2\sqrt{u_2}\right)\mathrm{erfc}\left(\sqrt{u_1}-\frac{\sqrt{u_2}}{\sqrt{u_1}}\right)\right.$$

$$\left.+\exp\left(2\sqrt{u_2}\right)\mathrm{erfc}\left(\sqrt{u_1}+\frac{\sqrt{u_2}}{\sqrt{u_1}}\right)\right]$$

(3.61)

For steady-state conditions, applying the method of images to Equation 3.56 yields

$$T(x,y,z)=T_0+\frac{q_{tb}}{4\pi\lambda_m}\exp\left[\frac{u_r x}{2D_t}\right]\left\{\int_0^H\frac{\exp[-u_r r'/(2D_t)]}{r'}-\int_0^H\frac{\exp[-u_r r'/(2D_t)]}{r'}\right\}dz_0$$

(3.62)

Introducing the dimensionless variables $R = r/H(L = H)$, $Z = z/H$, $Z' = z_0/H$, $U_1 = R'^2/(4Fo)$, $U_2 = Pe^2R'^2/16$, and $R'^2 = R^2 + (Z - Z')^2$, we can express Equations 3.60 and 3.62 in dimensionless forms as follows:

$$\Theta_{MFLS}=\exp\left[\frac{Pe}{2}R\cos\varphi\right]\left\{\int_0^1\frac{\Gamma(1/2,U_1;U_2)}{\sqrt{\pi}R'}\,dZ'-\int_{-1}^0\frac{\Gamma(1/2,U_1;U_2)}{\sqrt{\pi}R'}\,dZ'\right\}$$

(3.63)

and for steady-state conditions:

$$\Theta_{MFLS\_steady\text{-}state}=\exp\left[\frac{Pe}{2}R\cos\varphi\right]\left\{\int_0^1\frac{\exp[-Pe\,R'/2]}{R'}\,dZ'\right.$$

$$\left.-\int_{-1}^0\frac{\exp[-Pe\,R'/2]}{R'}\,dZ'\right\}$$

(3.64)

The mean temperature at the borehole wall for the **MFLS** for transient and steady-state conditions is as follows:

$$T(r_0,z) = T_0 + \frac{q_{tb}}{4\pi\lambda_m} \, \mathrm{I}_0\left[\frac{u_t r}{2D_t}\right] \left\{ \int_0^H \frac{\exp[-u_t r'/(2D_t)]}{r'} - \int_0^H \frac{\exp[-u_t r'/(2D_t)]}{r'} \right\} dz_0$$

(3.65)

$$T(r_0,z) = T_0 + \frac{q_{tb}}{4\pi\lambda_m} \, \mathrm{I}_0\left[\frac{u_t r}{2D_t}\right] \left\{ \int_0^H \frac{\exp[-u_t r'/(2D_t)]}{r'} - \int_{-H}^0 \frac{\exp[-u_t r'/(2D_t)]}{r'} \right\} dz_0$$

(3.66)

$$\Theta_{\mathrm{MFLS}} = \mathrm{I}_0\left[\frac{Pe}{2} R\right] \left\{ \int_0^1 \frac{\Gamma(1/2, U_1; U_2)}{\sqrt{\pi} R'} dZ_0 - \int_{-1}^0 \frac{\Gamma(1/2, U_1; U_2)}{\sqrt{\pi} R'} dZ_0 \right\}$$

(3.67)

$$\Theta_{\mathrm{MFLS\_steady\text{-}state}} = \mathrm{I}_0\left[\frac{Pe}{2} R\right] \left\{ \int_0^1 \frac{\exp[-Pe\,R'/2]}{R'} dZ_0 - \int_{-1}^0 \frac{\exp[-Pe\,R'/2]}{R'} dZ_0 \right\}$$

(3.68)

where $R' = \sqrt{R^2 + R_0^2 - 2RR_0 \cos\varphi_r' + (Z - Z')^2}$ and $R_0 = r_0/L$.

Figure 3.17 shows the difference between the MFLS and ILS models due to simulation time and borehole length. The longer the simulation time and the shorter the borehole length become, the larger the discrepancy is.

Molina-Giraldo et al. (2011b) concluded that the role of axial effects depends on the groundwater velocity and the length of the BHE. They stated that the MFLS can be applied to all groundwater flow conditions and borehole lengths. However, they also found that the FLS is still valid for Pe < 1.2 and the MILS for Pe > 10.

The coded functions (MATLAB scripts) of the MFLS model for transient (Equations 3.60 and 3.61) and steady-state (Equation 3.62) conditions are listed as T_MFLS.m and T_MFLSs.m, respectively. For computing the mean temperature at the borehole wall (Equations 3.65 and 3.66), the coded functions are T_MFLSc.m and T_MFLScs.m, respectively.

Figure 3.18 compares the relative temperature, ΔT, contours obtained by the MFLS and MILS models with groundwater advection (Molina-Giraldo

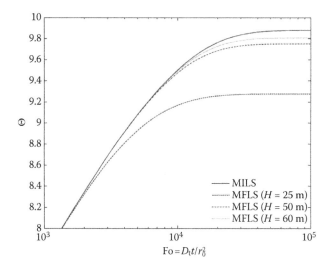

Figure 3.17 Temperature response over dimensionless number Fo for different bore-hole lengths ($q = 1 \times 10^{-7}$ m s^{-1}, $D_t = 9 \times 10^{-7}$ m^2 s^{-1}, $x = r_0 = 0.05$ m, $y = 0$ m, $z = 0.5 \times H$). (Modified after Molina-Giraldo, N. et al. *International Journal of Thermal Sciences* 50 (12), 2506–2513, 2011.)

et al. 2011b). Temperature plumes simulated by the MFLS model are shorter (Figure 3.18a). The reason for this is an axial effect. It induces vertical dissipation of heat and thus leads to lower temperature changes at any lateral distance from the BHE than the MILS. Consequently, by ignoring axial effects, longer boreholes are calculated for the same energy demand by the MILS compared to the MFLS (Marcotte et al. 2010). Figure 3.18b shows the axial extension of the temperature plume. It is revealed that the discrepancies between the MFLS and MILS models are most pronounced close to the endpoints of the borehole.

3.1.9 Infinite plane source— one-dimensional conduction

The **instantaneous plane source** of strength dE/dA (J m^{-2}), within an infinite porous medium is described by a one-dimensional heat conduction equation with constant coefficients, where x is the direction normal to the plane source, and A is the source area:

$$\frac{\partial T}{\partial t} = D_t \frac{\partial^2 T}{\partial x^2}$$

(3.69)

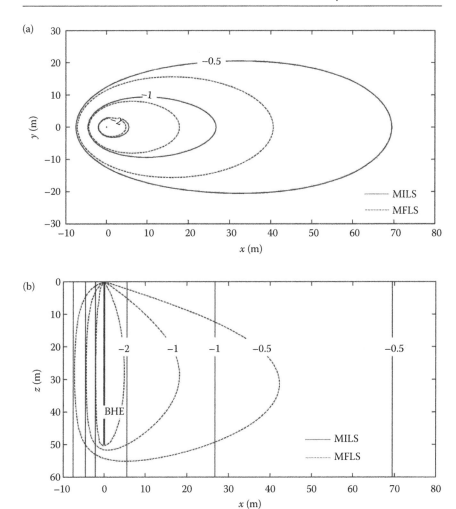

Figure 3.18 Temperature response (K) for a single BHE with constant energy extraction (q_{tb} = J/H = 20 W m^{-1}, q = 1.0 × 10^{-7} m/s, H = 10 m). (a) Plan view (z = 0.5 × H for MFLS). (b) Vertical cross section along centerline (y = 0 m). (Modified after Molina-Giraldo, N. et al. *International Journal of Thermal Sciences* 50 (12), 2506–2513, 2011.)

The initial condition is $T(x, t = 0) = T_0$. The temperature $T(x,t)$ for a source at the location x_0 is

$$T(x,t) = T_0 + \frac{\Delta E/A}{2 \cdot C_m \cdot (\pi D_t t)^{1/2}} \exp\left(-\frac{(x-x_0)^2}{4 D_t t}\right) \qquad (3.70)$$

3.1.10 Moving infinite plane source—
one-dimensional conduction and advection

The **continuous plane source** of specific strength j (W m^{-2}) within an infinite porous medium consists of a one-dimensional heat conduction and advection problem with the differential equation

$$\frac{\partial T}{\partial t} = D_t \frac{\partial^2 T}{\partial x^2} - u_{t,x} \frac{\partial T}{\partial x} \tag{3.71}$$

The initial condition is $T(x, t = 0) = T_0$. The temperature $T(x, t)$ for a source at the location x_0 is—after Bear (1979)—in analogy to solute transport:

$$T(x,t) = T_0 + \frac{\Delta T u_t}{(4\pi D_{t,L})^{1/2}} \exp\left(\frac{u_t \cdot (x - x_0)}{2D_{t,L}}\right) \int_0^t \frac{1}{\sqrt{\tau}} \exp\left(-\frac{a}{\tau} - b\tau\right) d\tau \tag{3.72}$$

where a and b are

$$a = \frac{(x - x_0)^2}{4D_{t,L}}; \quad b = \frac{u_t^2}{4D_{t,L}} \tag{3.73}$$

and using the steady-state temperature change ΔT:

$$\Delta T = \frac{j}{C_w q} \tag{3.74}$$

The integral may be evaluated numerically. Alternatively, the solution for the **continuous plane source** can be transformed as follows:

$$T(x,t) = T_0 + \frac{\Delta T}{2} \exp\left(\frac{u_t \cdot (x - x_0)}{2D_{t,L}}\right) \times \left[\exp\left(\frac{-u_t \cdot |x - x_0|}{2D_{t,L}}\right) \mathrm{erfc}\left(\frac{|x - x_0| - u_t t}{\sqrt{4D_{t,L}t}}\right)\right.$$

$$\left. - \exp\left(\frac{u_t \cdot |x - x_0|}{2D_{t,L}}\right) \mathrm{erfc}\left(\frac{|x - x_0| + u_t t}{\sqrt{4D_{t,L}t}}\right) \right] \tag{3.75}$$

Note that the products exp(arg1) × erfc(arg2) may have to be evaluated for very large arguments arg1 by combining the series expansion of the two related functions. The continuous plane source may correspond to a dense array of heat exchangers in the vertical plane of an aquifer with uniform flow conditions.

Dimensionless temperature profiles can be obtained by using the **scaled variables** x', T' and t':

$$x' = \frac{u_t \cdot (x - x_0)}{D_{t,L}}; \quad t' = \frac{u_t^2 t}{D_{t,L}}; \quad T' = \frac{T - T_0}{\Delta T} \tag{3.76}$$

An evaluation is shown in Figure 3.19. Note that **close to the source** at $x' = 0$, thermal **dispersion is overestimated** compared to field observations. This is due to the fact that the combined effect of grain scale mechanical dispersion and macrodispersion depends on the flow distance (Section 2.1.2.3). Both mechanical and macrodispersion start from zero at the source. Therefore, they need some time and flow distance, respectively, until a constant dispersion coefficient is attained. Furthermore, for small times ($t' < 1$), the temperature distribution is almost symmetrical due to the dominance of diffusion and dispersion. For **large times after start of injection** ($t' > 100$), the heat propagation is strongly influenced by advection and can be approximated by

$$T(x,t) \cong T_0 + \frac{\Delta T}{2} \operatorname{erfc}\left(\frac{x' - t'}{2\sqrt{t'}}\right) = T_0 + \frac{\Delta T}{2} \operatorname{erfc}\left(\frac{x - u_t t}{2\sqrt{D_{t,L} t}}\right) \tag{3.77}$$

This solution corresponds to the development of an initially sharp temperature front in an infinite porous medium with $T(x < 0, t = 0)$ and $T(x > 0, t = 0) = T_0 + \Delta T$.

MATLAB scripts of the **infinite plane source** model for transient conductive/dispersive–advective transport (Equation 3.75) and of the **development**

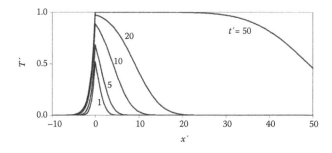

Figure 3.19 Continuous plane source in one-dimensional aquifer. Dimensionless results with $x' = x u_t / D_{t,L}$; $t' = t u_t^2 / D_{t,L}$; $T' = T/T_0$.

of an initially sharp temperature front (Equation 3.77) in a uniform flow field are listed as *Continuous_injection.m* and *Thermal_front.m*, respectively.

3.1.11 Steady-state injection into an aquifer with thermally leaky top layer

The **steady-state, one-dimensional heat transport with thermally leaky top layer** (overburden or cap rock) in a shallow aquifer is described according to Equation 2.113:

$$D_{t,L} \frac{\partial^2 T}{\partial x^2} - u_{t,x} \frac{\partial T}{\partial x} + \frac{j_{vert,top}}{mC_m} = 0 \tag{3.78}$$

where the **thermal flux from the surface to the groundwater** $j_{vert,top}$, considering only heat conduction and assuming constant surface temperature $T_{surface}$, is

$$j_{vert,top} = \frac{\lambda_{vert} \cdot (T_{surface} - T)}{(f + m/2)} \tag{3.79}$$

The **temperature profile** $T(x)$ is the vertically averaged value within the thin aquifer. The temperature before the thermal injection is T_0. The bottom layer is assumed to behave like an insulator. The boundary condition is $T(x \rightarrow \infty) = 0$. The solution for $x \geq 0$ (after Bear 1979) is in analogy to solute transport:

$$T(x) = T_{surface} + \frac{\Delta T}{\chi} \exp\left(\frac{u_t \cdot (x - x_0)}{2D_{t,L}} \cdot (1 - \chi) \right) \tag{3.80}$$

with

$$\chi = \left(1 + \frac{4D_{t,L}\lambda_{vert}}{C_m m \cdot (f + m/2)u_t^2} \right)^{1/2} \tag{3.81}$$

If the expression

$$\frac{4D_{t,L}\lambda_{vert}}{C_m m \cdot (f + m/2)u_t^2} \tag{3.82}$$

within Equation 3.81 is small, which is often the case, then $1/\chi \cong 1$, and, using the first term of a series expansion for χ, $(1 - \chi)$ is

$$1 - \chi \simeq \frac{2D_{t,L}\lambda_{vert}}{C_m m \cdot (f + m/2)u_t^2} \tag{3.83}$$

Thus, the **temperature profile is approximately**

$$T(x) \simeq T_{surface} + \Delta T \exp\left(\frac{(x - x_0)\lambda_{vert}}{C_m m \cdot (f + m/2)u_t}\right) = T_{surface}$$

$$+ \Delta T \exp\left(\frac{(x - x_0)\lambda_{vert}}{C_w m \cdot (f + m/2)q}\right) \tag{3.84}$$

In this case, the steady-state temperature profile is independent of the coefficient $D_{t,L}$. Therefore, it cannot be used to estimate $D_{t,L}$ based on data along a flow line. Instead the profile is determined by the heat flow in the overburden between surface and aquifer. It can be employed to assess the mitigation effect of the overburden for thermal use.

As an **example**, consider the parameter values $D_{t,L} = 3$ m^2 day^{-1}, $m = 10$ m, $f = 5$ m, $q = 1$ m day^{-1}, $\lambda_{vert} = 0.0015$ kJ m^{-1} K^{-1} s^{-1}, $C_w = 4200$ kJ m^{-3} K^{-1}, $C_m = 3000$ kJ m^{-3} K^{-1}, and $x_0 = 0$ m. Expression 3.82 is about 0.003, and the approximation 3.84 is well applicable. The temperature at $x = 100$ m is about $T_0 + \Delta T \times 0.97$. For $(T - T_0)/\Delta T = 0.5$, which corresponds to an increase or decrease in the temperature by $\Delta T \times 0.5$ (half-value distance), the flow distance needed is about $x \cong 2200$ m.

3.1.12 Harmonic temperature boundary condition for one-dimensional conductive–advective heat transport

3.1.12.1 One-dimensional vertical conductive heat transport

One-dimensional vertical thermally diffusive transport caused by a harmonic fluctuation of the surface temperature, superimposed by a geothermal gradient, is given by the extended classical equation of the periodic temperature profile below the soil surface (Gröber et al. 1955) assuming constant coefficients:

$$T(z,t) = T_{surface} - Gz + \Delta T \exp\left(z\sqrt{\frac{\pi}{D_t t_p}}\right)\cos\left(z\sqrt{\frac{\pi}{D_t t_p}} + 2\pi\frac{t}{t_p}\right) \tag{3.85}$$

Note that at soil surface, $z = 0$, and with increasing soil depth, $z < 0$. The symbol $T_{surface}$ (K) is the mean soil surface temperature, ΔT (K) is the amplitude of the temperature fluctuation at the surface, G (K m^{-1}) is the geothermal gradient, D_t (m^2 s^{-1}) is the thermal diffusion coefficient of the soil, and t_p (s) is the period of the harmonic temperature development ($t_p = 365.25$ days for seasonal fluctuations).

3.1.12.2 One-dimensional horizontal conductive/ dispersive–advective transport

Let us consider a **horizontal aquifer layer** with uniform flow in the x-direction. Top and bottom layers are assumed to act as **insulators**. One-dimensional horizontal diffusive/dispersive and advective **heat transport** is described by the differential equation

$$\frac{\partial T}{\partial t} = D_{t,L}\frac{\partial^2 T}{\partial x^2} - u_{t,x}\frac{\partial T}{\partial x} \tag{3.86}$$

At location $x = 0$, the **prescribed harmonic temperature boundary condition** applies, for example, as the result from river water inflow with seasonal temperature fluctuation:

$$T(x = 0, t) = T_0 + \Delta T\cos(\omega t) \tag{3.87}$$

The second boundary condition is $T(x \to \infty) = 0$. The coefficient ω is the angular frequency with

$$\omega = \frac{2\pi}{t_p} \tag{3.88}$$

where t_p is the period. The **solution** is of the form (after Burger et al. 1984)

$$T(x, t) = T_0 + \Delta T\exp(-ax)\cos(-bx + \omega t) \tag{3.89}$$

The **coefficients** a and b are the attenuation coefficient and the wave number, respectively. In order to fulfill the differential equation and its boundary conditions, the coefficients are chosen as follows:

$$b = \sqrt{\frac{-\dfrac{u_t^2}{4D_{t,L}} + \sqrt{\left(\dfrac{u_t^2}{4D_{t,L}}\right)^2 + \omega^2}}{2D_{t,L}}} \; ; \quad a = \frac{\omega - bu_t}{2bD_{t,L}} \tag{3.90}$$

One-dimensional harmonic horizontal conductive/dispersive–advective transport may correspond to infiltration of river water into an aquifer or to a dense array of heat exchangers with harmonic temperature variation in the vertical plane of an aquifer with uniform flow conditions.

A **scaled evaluation** for the period $t_p = D_{t,L}/u_t^2$ is shown in Figure 3.20. It visualizes the damping effect of thermal conduction and macrodispersion for the selected angular frequency.

If, based on experimental data, the coefficients a and b can be evaluated, the **parameters** u_t and $D_{t,L}$ can be determined as follows:

$$D_{t,L} = \frac{a\omega}{b \cdot (a^2 + b^2)}; \quad u_t = \frac{\omega \cdot (a^2 - b^2)}{b \cdot (a^2 + b^2)} \tag{3.91}$$

As an **illustration**, we select the measured temperature data of Trüeb (1976). He measured temperature time series within a borehole located at a distance of 500 m from the infiltrating River Rhine north of Zurich. Measurements were performed before and after the construction of the dam Rheinau (Figure 3.21), which completely changed the groundwater flow regime after 1957, but not so much the flow direction. From the time delay $\Delta t \cong 180$ day, the coefficient b can be evaluated for the situation before existence of the dam, yielding the angular frequency $\omega = 6.28$ year^{-1} = 0.0172 day^{-1} to be $b \cong 6.2 \times 10^{-3}$ m^{-1}. The temperature amplitude of the river water was $\Delta T_0 \cong 8.3$ K, and in the borehole, an amplitude of $\Delta T_1 \cong 1.5$ K was

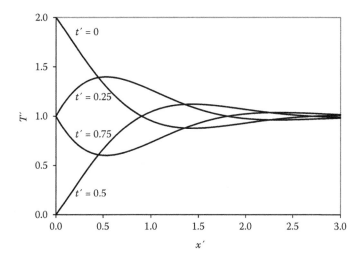

Figure 3.20 Harmonic boundary condition in one-dimensional semi-infinite aquifer. Dimensionless results with $x' = xu_t/D_{t,L}$; $t' = tu_t^2/D_{t,L}$; $T' = T/T_0$. The period is selected according to $u_t^2/D_{t,L} = 1$.

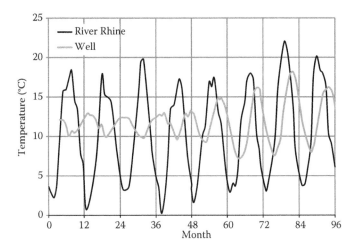

Figure 3.21 Measured temperature of Rhine River and groundwater observation well (500 m from infiltration) before and after the construction of the Rheinau dam (Switzerland) after Trüeb (1976). Month 0 = January 1953: filling from 1957 on (about month 48).

observed. Therefore, the coefficient a is evaluated as $a \cong 3.4 \times 10^{-3}$ m^{-1}, and the coefficients $D_{t,L} \cong 190$ m^2 day^{-1} and $u_t \cong 1.5$ m day^{-1} result from using Equation 3.91. The longitudinal thermal macrodispersivity would be $\beta_L \cong 128$ m. Thermal diffusion can be neglected. For the situation after construction of the dam, the time delay was $\Delta t \cong 84$ day, and the ratio $\Delta T_1/\Delta T_1 \cong 4.5/8.3$. Further, $D_{t,L} \cong 740$ m^2 day^{-1} and $u_t \cong 4.1$ m day^{-1}. The corresponding longitudinal thermal macrodispersivity in that case is $\beta_L \cong 179$ m. The longitudinal thermal dispersivity is roughly constant for both cases. However, we have to keep in mind that the harmonic model assumes that top and bottom layers act as insulators.

Equation 3.89 can also be applied to the vertical case as described in Section 3.1.12.1, including the assumption of constant vertical thermal advection due to uniform recharge.

A MATLAB script of the model for one-dimensional **harmonic thermal conductive/dispersive–advective transport** (Equations 3.89 and 3.90) is listed as *Harmonic_temperature.m*.

3.1.12.3 Horizontal layer embedded in conductive bottom and top layer

An **analytical approximation** can be formulated by taking the effect of thermally conductive bottom and top layers into account. The prerequisite is that both layers are homogeneous and are both sufficiently thick. For example, in order to exclude an influence of the soil surface, the distance from soil surface to groundwater table has to be large enough. The one-dimensional **transient differential equation** for the horizontal aquifer layer reads as follows:

$$\frac{\partial T}{\partial t} = D_{t,L}\frac{\partial^2 T}{\partial x^2} - u_{t,x}\frac{\partial T}{\partial x} + \frac{j_{vert,top}}{C_m m} + \frac{j_{vert,bottom}}{C_m m} \tag{3.92}$$

with

$$j_{vert,top} = -\lambda_{top}\left(\frac{\partial T}{\partial z}\right)_{y=0}; \quad j_{vert,bottom} = \lambda_{bottom}\left(\frac{\partial T}{\partial z}\right)_{y=0} \tag{3.93}$$

The concept for the horizontal temperature development in the aquifer layer is still the same as in Equation 3.89. For the **vertical temperature fluctuation**, the following approach is taken:

$$T(x, y, t) = T_0 + \Delta T \exp(-ax - a_0 y)\cos(-bx - b_0 y + \omega t) \tag{3.94}$$

with

$$a_0 = b_0; \quad b_0 = \sqrt{\frac{\omega}{2D_0}};$$

$$\lambda_0 = \lambda_{top} = \lambda_{bottom} = D_0 C_0;$$

$$D_0 = D_{top} = D_{bottom}; \tag{3.95}$$

$$C_0 = C_{top} = C_{bottom};$$

The vertical coordinate y_0 starts at the upper and lower aquifer layer boundary. In order to fulfill the differential equation and its boundary conditions, the **coefficients** a and b are chosen as follows:

$$b = \sqrt{\frac{-\left(\frac{u_t^2}{4D_{t,L}} + \eta\right) + \sqrt{\left(\frac{u_t^2}{4D_{t,L}} + \eta\right)^2 + (\omega + \eta)^2}}{2D_{t,L}}};$$

$$a = \frac{\omega + \eta - bu_t}{2bD_{t,L}}; \tag{3.96}$$

$$\eta = \sqrt{2\omega D_0}\,\frac{C_0}{C_m m}$$

Given the coefficients a and b, the **parameters** $D_{t,L}$ and u_t are evaluated as follows:

$$D_{t,L} = \frac{\omega - \sqrt{2\omega D_0}\ \dfrac{C_0}{C_m m}\left(\dfrac{b}{a}-1\right)}{\dfrac{b}{a}\cdot(a^2+b^2)};$$

$$u_t = \frac{D\cdot(a^2-b^2)+\sqrt{2\omega D_0}\ \dfrac{C_0}{C_m m}}{a}$$

(3.97)

For $D_0 = 0$, Equation 3.91 is obtained.

The **illustrative example** of Section 3.1.12.2 with the infiltration of river water and the estimation of the parameters before and after the construction of the dam in the Rhine River can thus be reevaluated. Using the same coefficients a and b, the parameters $D_{t,L}$ and u_t are obtained as follows, assuming $C_0 = 1.6 \times 10^6$ J m^{-3} K^{-1}, $C_m = 2.4 \times 10^6$ J m^{-3} K^{-1}, and $D_0 = 0.0972$ m^2 day^{-1}. Before construction of the dam, the parameters were $D_{t,L} \cong 167$ m^2 day^{-1}, $u_t \cong 1.3$ m day^{-1}, and $\beta_L \cong 128$ m. After construction of the dam, they were $D_{t,L} \cong 614$ m^2 day^{-1}, and $u_t \cong 3.4$ m day^{-1}, and $\beta_L \cong 178$ m. The comparison with Section 3.1.12.2 shows that the top and bottom layers may show a distinct influence on the parameters. However, interestingly, the values for longitudinal thermal macrodispersivity did not change with the new model. Again, very roughly, it is about constant for both cases. It has to be mentioned that the sensitivity of the parameters with respect to the values of a and b is relatively large.

3.2 OPEN SYSTEMS

Open systems in shallow aquifers often consist of one or several extraction wells and facilities for the injection into more or less uniform flow fields, which can be modeled as local infiltration wells. An overview on available analytical solutions for open systems is given in Table 3.1. This table shows that heat conduction is treated differently depending on the analytical solution. The analytical approach presented by Guimerà et al. (2007), for instance, does not account for heat transfer into the confining layers. On the other hand, some analytical approaches do not account for heat conduction within the aquifer (Lauwerier 1955; Malofeev 1960; Yang and Yeh 2008). In the following, we present some analytical solutions to flow and heat transport problems.

Table 3.1 Available analytical solutions for open systems

Source	Flow type			Heat conduction in aquifer		Heat conduction in confining layers		Top boundary condition	
	Linear	Radial	Arbitrary	Linear	Radial	Caprock	Bedrock	Variable temperature	Heat exchange
Lauwerier (1955)	x					inf	inf		
Malofeev (1960)		x				inf	inf		
Avdonin (1964)	x			x		inf	inf		
Avdonin (1964)		x			x	inf	inf		
Guimerà et al. (2007)		x			x	–	–		
Chen and Reddell (1983)		x			x	inf	inf		
Voigt and Haefner (1987)		x			x	fin	inf	x	
Güven et al. (1983)		x			x	fin	inf		
Yang and Yeh (2008)		x				fin	fin		x
Gringarten and Sauty (1975)			x			inf	inf		

Note: inf: infinite extent; fin: finite extent (constant aquifer thickness).

3.2.1 Analytical solution for steady-state flow in multiple well systems

Open systems are often operated at quasi–steady state. Therefore, **steady-state flow in multiple well systems of idealized two-dimensional aquifers with uniform flow** is of high interest. In this section, we would like to recall results of analytical solutions for the computation of isolines for **piezometric head** and **stream function** as a result of local sources and sinks (constant rate recharging and/or pumping wells, single wells, or series of wells) in a uniform horizontal flow field of an **infinite and confined aquifer without areal recharge**. The hydraulic conductivity K_w, as well as the thickness m of the aquifer, are assumed to be constant. The computation is performed analytically, and graphs of isolines of both the head and the stream function can be produced. The results are also approximately valid for phreatic aquifers, provided that the rise and decline of the groundwater table are small compared to the thickness of the aquifer.

The computation is performed in a homogeneous horizontal x–y system making use of the **superposition principle**. According to the potential flow theory, the specific discharge vector can be expressed by

$$q = -K_w \nabla h_w = -\nabla \varphi \tag{3.98}$$

where φ is the velocity potential with $\varphi = K_w h_w$. For a **uniform flow field**, the velocity potential is given by

$$\varphi(x, y) = -q_0 \cdot (x\cos\alpha + y\sin\alpha) \tag{3.99}$$

q_0 is the specific flux of the regional flow field, and α is the flow direction with respect to the x-axis (Figure 3.22). The stream function $\psi(x, y)$ of the uniform flow field is

$$\psi(x, y) = -q_0 \cdot (y\cos\alpha + x\sin\alpha) \tag{3.100}$$

For a **single well** (Figure 3.23), the velocity potential is (Bear 1979)

$$\varphi(x, y) = \frac{Q/m}{4\pi} \ln\left(\frac{(x - x_w)^2 + (y - y_w)^2}{R_w^2} \right) \tag{3.101}$$

where Q is the recharge or pumping rate ($Q > 0$: recharge; $Q < 0$: pumping); x_w and y_w are the coordinates of the well. R_w is the radius of influence of the well, and m is the aquifer thickness. An influence of the finite well

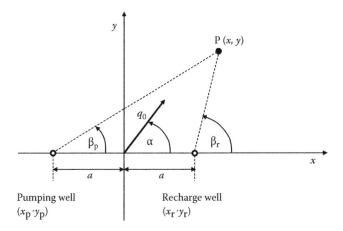

Figure 3.22 Double well system with recharge well and pumping well in uniform flow field q_0.

radius r_w on the flow is disregarded. The stream function of a single well (Figure 3.23) is

$$\psi(x,y) = \frac{Q/m}{2\pi} \tan^{-1}\left(\frac{y - y_w}{x - x_w}\right) \tag{3.102}$$

The computation of the head around a single well is based on the concept of a finite radius of influence of the well. According to this concept, it is

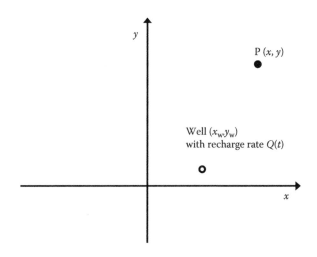

Figure 3.23 Well with recharge rate Q_w and observation point P.

assumed that the decline or rise of the head is zero at a distance of R_w. The result is only approximately valid inside of the zone of influence. Outside of the zone, results are inaccurate. In the case of a **two-well system** (Figure 3.22) with pumping and recharging wells of opposite recharge and pumping rate $\pm Q$ and same well radius r_w, the result of the superposition is valid in the whole domain regardless of the value of R_w (also outside of the zone of influence). The result is then independent of the radius of R_w. The same result is obtained by a consideration of a straight constant head line (e.g., a river or lake) by applying the **method of images** (introduction of a fictitious image well with opposite rate). In general, the solution becomes independent of the radius R_w if the pumping and infiltration rates of all wells sum up to zero.

An example of the flow field for a system with one pumping well and two injection wells with $Q_p = -2Q_R$ in a uniform flow field aligned with the direction defined by the pumping well and the central point between the two recharging wells is shown in Figure 3.24 in scaled form. The length scale is the half-distance a between the pumping well and the center between the recharge wells. Regional flow with specific discharge q_0 is from left to right. The pumping rate is chosen as $Q_P = \pi m\, a q_0$. It can be

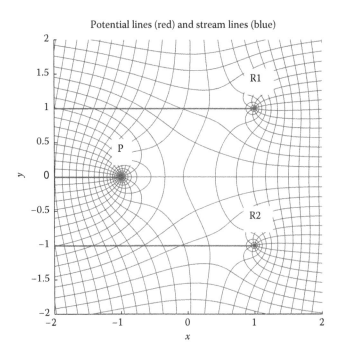

Potential lines (red) and stream lines (blue)

Figure 3.24 **(See color insert.)** Scaled flow field with $\alpha = 0$ and scaled pumping rate; one pumping well and two injection wells. P: pumping well; R1, R2: recharge wells.

seen that downstream of the pumping well, a stagnation point appears. Two further stagnation points are located upstream of the recharge wells. Furthermore, it can be seen that stream tubes from the regional flow pass between the wells. Note that the thick blue line in Figure 3.24 is due to the discontinuity of the stream function caused by the wells. The discontinuity is not necessarily a streamline.

3.2.1.1 Double well system in uniform flow field

The velocity potential of a double well system with pumping and recharging well in a uniform flow field with direction α (Figure 3.22), where both the recharge well and the pumping well have the same constant discharge rate Q (>0), is (DaCosta and Bennett 1960)

$$\varphi(x,y) = -q_0 \cdot (x\cos\alpha + y\sin\alpha) + \frac{Q/m}{4\pi}\ln\left(\frac{(x+a)^2 + y^2}{(x+a)^2 + y^2}\right) \tag{3.103}$$

The symbol a is the half-distance between the wells. The corresponding stream function is (DaCosta and Bennett 1960)

$$\psi(x,y) = -q_0 \cdot (y\cos\alpha - x\sin\alpha) + \frac{Q/m}{2\pi}(\theta_r - \theta_p)$$

$$= -q_0 \cdot (y\cos\alpha - x\sin\alpha) + \frac{Q/m}{2\pi}\tan^{-1}\left(\frac{2ay}{a^2 - x^2 - y^2}\right) \tag{3.104}$$

By introducing the dimensionless pumping rate χ with

$$\chi = \frac{Q}{\pi m a q_0} \tag{3.105}$$

and dividing Equation 3.104 by the pumping rate, we obtain

$$\frac{\psi(x,y)}{Q/m} = -\frac{1}{\pi\chi}\cdot\left(\frac{y}{a}\cos\alpha - \frac{x}{a}\sin\alpha\right) + \frac{1}{2\pi}(\theta_p - \theta_r) \tag{3.106}$$

For flow parallel to the x-axis (angle $\alpha = 0$), the flow field is shown in Figure 3.25 for the dimensionless pumping rates $\beta = 0.5$ (weak discharge rate; Figure 3.25a), $\beta = 1$ (Figure 3.25b), and $\beta = 2$ (strong discharge rate; Figure 3.25c). By analyzing these flow fields, it can be seen that for $\beta = 0.5$ (Figure 3.25a), the limiting streamlines of the wells, which define the well

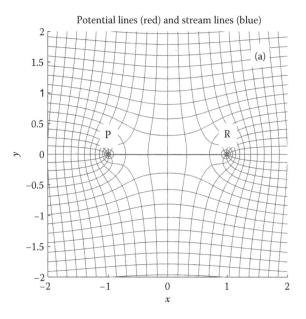

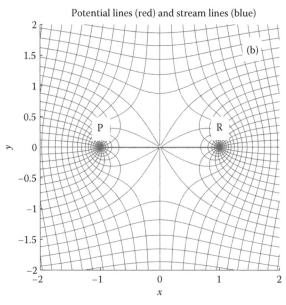

Figure 3.25 **(See color insert.)** (a) Scaled flow field with $\alpha = 0$ and scaled pumping rate $\chi = 0.5$. P: pumping well; R: recharge well. (b) Scaled flow field with $\alpha = 0$ and $\chi = 1$. (c) Scaled flow field with $\alpha = 0$ and $\chi = 2$. (d) Scaled flow field with $\alpha = 90°$ and $\chi = 1$.

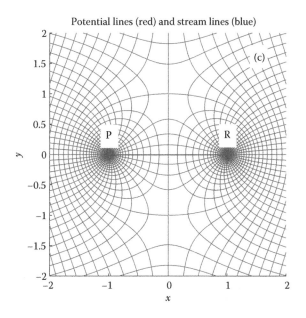

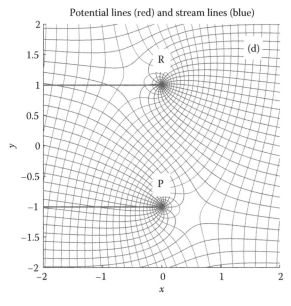

Figure 3.25 (Continued) (a) Scaled flow field with $\alpha = 0$ and scaled pumping rate $\chi = 0.5$. P: pumping well; R: recharge well. (b) Scaled flow field with $\alpha = 0$ and $\chi = 1$. (c) Scaled flow field with $\alpha = 0$ and $\chi = 2$. (d) Scaled flow field with $\alpha = 90°$ and $\chi = 1$.

recharge and discharge regions, are well separated from each other. Part of the region between the wells is flushed by regional flow. This situation is characterized by two stagnation points along the x-axis. For β = 2 (Figure 3.25c), the limiting streamlines of the wells intersect and show a distinct recirculation between the wells. This effect is visible by the fact that one or more stream tubes start at the recharge well and end at the pumping well. In this case, two stagnation points appear along the y-axis. For the critical case with β = 1 (Figure 3.25b), the limiting streamlines join at point x = 0, y = 0, which is the only stagnation point. The three cases can also be characterized by analyzing the location of the stagnation points. In general, two stagnation points show up. At a **stagnation point**, by definition, the components q_x and q_y of the discharge vector vanish, that is (for α = 0),

$$q_x(x_s, y_s) = q_0 + \frac{Q}{2\pi m} \cdot \left[-\frac{(a - x_s)}{(a - x_s)^2 + y_s^2} - \frac{(a + x_s)}{(a + x_s)^2 + y_s^2} \right] = 0$$

$$q_y(x_s, y_s) = \frac{Q}{2\pi m} \cdot \left[-\frac{y_s}{(a - x_s)^2 + y_s^2} - \frac{y_s}{(a + x_s)^2 + y_s^2} \right] = 0$$

(3.107)

These two equations can be rearranged as follows:

$$\frac{Qa}{\pi m q_0} \cdot \left[-\left(a^2 - x_s^2\right) - y_s^2 \right] + \left[\left(a^2 + x_s^2 + y_s^2\right)^2 - 4a^2 x_s^2 \right] = 0$$

(3.108)

$$\frac{Qa}{\pi m q_0} 2x_s y_s = 0$$

For this system, two solutions exist, one for $x_s = 0$, and one for $y_s = 0$. For $x_s = 0$ (for α = 0)

$$y_s = \pm a \sqrt{\frac{Q}{\pi m a q_0} - 1} = \pm a \sqrt{\chi - 1}$$

(3.109)

which is meaningful for $\chi \geq 1$. This situation corresponds to Figure 3.25c. Equation 3.109 defines two stagnation points in general.

For the other case with $y_s = 0$ (for α = 0)

$$x_s = \pm a \sqrt{1 - \frac{Q}{\pi m a q_0}} = \pm a \sqrt{1 - \chi}$$

(3.110)

which is meaningful for $\chi \leq 1$. This situation corresponds to Figure 3.25a. For the **critical case** with $\chi = 1$ (Figure 3.25b), $x_s = 0$ and $y_s = 0$. In this case, the two stagnation points merge to one at the origin. This is the case with minimum distance between the wells, where no recirculation between the wells occurs. Recirculation between the wells can be analyzed by evaluating the stream function passing through one stagnation point and the two wells and the stream function passing through the origin (DaCosta and Bennet 1960). Taking the difference yields half the recirculation rate. The total recirculation rate I can be expressed by using information on stagnation point S_1:

$$I = 2 \cdot [\psi(x_{s1}, y_{s1}) - \psi(0, 0^+)] = 2 \cdot \left[\psi(x_{s1}, y_{s1}) + \frac{Q}{2m} \right] \tag{3.111}$$

or with information of stagnation point S_2:

$$I = 2 \cdot [\psi(0, 0^-) - \psi(x_{s2}, y_{s2})] = 2 \cdot \left[\frac{Q}{2m} - \psi(x_{s2}, y_{s2}) \right] \tag{3.112}$$

The recirculation rate denotes the rate that stems from the recharge well. Note the singularity of the stream function at the origin. Recirculation is characterized by a positive value. Negative values are an indication for the amount of regional flow between the wells. In such a case, no recirculation occurs. After insertion of the coordinates of one stagnation point, the recirculation rate divided by the discharge rate Q is for $\chi \geq 1$ and $\alpha = 0$ (Bear 1979):

$$\frac{I}{Q} = \frac{2}{\pi} \left[-\frac{1}{\chi} \sqrt{\chi - 1} + \tan^{-1} \left(\sqrt{\chi - 1} \right) \right] \tag{3.113}$$

The **recirculation rate** I between the wells is zero for $\chi = 1$, which has already been referred to as **critical case**. The corresponding half-distance a between the wells is for $\alpha = 0$:

$$a = \frac{Q}{\pi m q_0} \tag{3.114}$$

The related distance between the wells is $d = 2a$. Equation 3.114 represents the **basis for the design of double well systems in parallel flow, aligned with the direction defined by the two wells** ($\alpha = 0$). Therefore, it has been used in the past in open systems for thermal use of shallow aquifers in order to avoid recirculation and thus avoid pumping of cool or warm water stemming from the injection well. Note that the considerations are based on pure advection, thus neglecting thermal diffusion and transverse dispersion effects.

For the case where the **flow field is not aligned with the direction defined by the two wells**, that is, $\alpha \neq 0$, the **location of the stagnation points** can be found best by applying the theory of complex numbers (DaCosta and Bennett 1960):

$$\frac{x_s}{a} = \mp\sqrt{\frac{1}{2}\left[1 - \chi\cos\alpha + \sqrt{1 + \chi^2 - 2\chi\cos\alpha}\right]}$$

$$\frac{y_s}{a} = \pm\sqrt{\frac{1}{2}\left[\chi\cos\alpha - 1 + \sqrt{1 + \chi^2 - 2\chi\cos\alpha}\right]}$$

(3.115)

The permissible pairs of x_s and y_s are those of opposite sign. Recirculation between wells can be evaluated in a similar manner as stated above (Equations 3.111 and 3.112) by using the stream function at the stagnation points and the origin.

Evaluations of the recirculation by DaCosta and Bennet (1960) for **different angles** α indicate that for the critical case with $\chi = 1$, there is **no recirculation** for the angles $-101° \leq \alpha \leq 101°$. They further showed that the angle α_{min} for minimum recirculation is

$$\cos\alpha_{min} = \frac{Q}{2\pi a q_0} = \frac{\chi}{2}$$

(3.116)

For $\alpha = 60°$, the amount of regional flow is maximum with $I/Q = -0.14$. This means that the design of the two-well system using Equation 3.114 is quite robust with respect to variations in the angle α of the regional flow field. For the ratio $\chi = 4/\pi = 1.273$, recirculation between the wells occurs except at the angle $\alpha = 50.46°$. This means that, theoretically, solutions without recirculation exist, where the half-distance a between the wells can be chosen to be smaller than the value in Equation 3.114. The corresponding value for the half-distance a would be

$$a_{min} = \frac{Q}{4mq_0}$$

(3.117)

as stated, for example, by Mehlhorn et al. (1981). However, if recirculation has to be prevented, this situation is not recommended due to the variability and uncertainty of the angle α of the flow field in practice.

An example with $\alpha = 90°$ and $\chi = 1$ is shown in Figure 3.25d. As can be seen, stream tubes from regional flow pass between the wells. Therefore, it is obvious that no recirculation occurs, as already expected from the discussion above. The related regional flow compared with Q is about -0.06 (no recirculation).

A **program** (MATLAB script) to visualize steady-state flow nets (head and stream function isolines) for multiple well systems in a uniform flow field is *wells_in_flow_field.m*. Note that the stream function for wells shows a distinct singularity (step). It is plotted as a thick blue line. Streamlines across this singularity line are not necessarily continuous. Often, continuity of streamlines can be achieved by proper choice of minimum and step values of the stream function contours.

The **coded function** (MATLAB script) for calculating the recirculation rate and the fraction of recirculation at the total pumping rate Q of double well systems is listed as *recirculation_rate.m*. For example, wells of a half-distance $a = 100$ m, which are aligned with the flow field at the angle $\alpha = 0$, with the pumping rate $Q = 4000$ m³ day^{-1} in an aquifer of thickness $m = 10$ m, and a Darcy velocity $q_0 = 1$ m day^{-1}, yield a recirculation rate of $I = 18.1$ m³ day^{-1} and a fraction of 0.0045 of water from the recharge well.

Lippmann and Tsang (1980) addressed the **problem of advective thermal breakthrough time at the pumping well** of a double well system, where the **flow is aligned with the direction between infiltration and pumping well** ($\alpha = 0$, Figure 3.22). The breakthrough time is the time that is needed for the thermal front after starting a double well system to reach the pumping well. Thermal diffusion and dispersion are disregarded.

For the situation where **no regional flow** exists (i.e., $q_0 = 0$), the **breakthrough time** t_b is (Lippmann and Tsang 1980)

$$t_b = \frac{\pi\phi m d^2}{3Q} \cdot \frac{C_m}{\phi C_w} = \frac{\pi\phi m d^2}{3Q} \cdot R_{t\_ret} \tag{3.118}$$

where R_{t_ret} is the thermal retardation factor after Equation 2.98, $d = 2a$ is the distance between the wells, and Q is the pumping and infiltration rate. Lippmann and Tsang (1980) also provide an analytic approximation for the temperature development $T(t)$ at the pumping well.

For the case where between pumping well and injection well a **flow field** $q_0 > 0$ exists, the **breakthrough time** is determined by (Lippmann and Tsang 1980)

$$t_b = \frac{d R_t \phi}{q_0} \cdot \left[1 - \frac{4A}{\sqrt{4A-1}} \tan^{-1}\left(\frac{1}{\sqrt{4A-1}} \right) \right] \tag{3.119}$$

where the auxiliary variable A is

$$A = \frac{Q}{2\pi m d q_0} \tag{3.120}$$

provided that the specific flow rate of the flow field q_0 is smaller than that of the critical case according to Equation 3.114, that is,

$$q_0 < \frac{2Q}{\pi m d} \tag{3.121}$$

For larger flow rates, the breakthrough time is infinite or does not exist. The latter is also applicable for $q_0 < 0$, that is, a flow direction from injection well to pumping well.

3.2.2 Linear flow

Pioneering work by Lauwerier (1955) presented an analytical solution in which the flow pattern is linear. Heat conduction is considered in the vertical direction toward the confining layers, which are assumed infinite in the z-direction (Figure 3.26). The analytical solution is as follows:

$$\bar{T}(\tau,\xi,\eta) = \mathrm{erfc}\left[\frac{\xi + |\eta| - 1}{2\sqrt{\theta(\tau-\xi)}}\right] U(\tau-\xi) \tag{3.122}$$

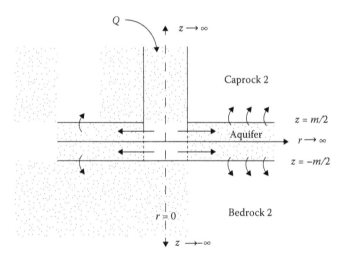

Figure 3.26 Conceptual model of Lauwerier (1955) analytical solution. (Modified after Voigt, H.D. and Haefner, F., *Water Resources Research* 23 (12), 2286–2292, 1987.)

subject to the following condition:

$$U(\tau - \xi) = 0 \quad \text{if } \tau - \xi \leq 0$$
$$U(\tau - \xi) = 1 \quad \text{if } \tau - \xi > 0 \tag{3.123}$$

The quantities $\bar{T}$, ξ, η, τ, and θ are dimensionless variables defined as

$$\bar{T} = \frac{T - T_0}{T_{inj} - T_0}, \quad \theta = \frac{C_m}{C_{m2}}, \quad \tau = \frac{4\lambda_{m2}t}{m^2 C_m}, \quad \xi = \frac{4\lambda_{m2}x}{m^2 C_w q_0} \tag{3.124}$$

$$\eta = \frac{2z}{H} \quad \text{for } z \geq \frac{m}{2} \tag{3.125}$$

where T_0 denotes the initial temperature of the aquifer, T_{inj} is the injection temperature, and C_{m2} and λ_{m2} are the volumetric heat capacity and thermal conductivity of the confining layers, respectively.

Avdonin (1964) gets rid of one of the restrictions from Lauwerier (1955) by adding thermal conduction also in the flow direction (x-direction). The analytical solution is as follows:

$$\bar{T} = \frac{\bar{x}}{\sqrt{\pi \lambda_e \tau}} \int_0^1 \exp\left[-\left(\gamma\sqrt{\lambda_e \tau}\psi - \frac{\bar{x}}{2\sqrt{\lambda_e \tau \psi}}\right)^2\right] \text{erfc}\left(\frac{\sqrt{\tau/\lambda_e}\psi^2}{2a\sqrt{1-\psi^2}}\right)\frac{d\psi}{\psi^2} \tag{3.126}$$

where:

$$\lambda_e = \lambda_m/\lambda_{m2}; \quad \gamma = \frac{mq_0 C_w}{4\lambda_m}; \quad \bar{x} = \frac{2x}{m}; \quad a = \sqrt{\frac{\lambda_{m2}C_m}{\lambda_m C_{m2}}} \tag{3.127}$$

Further details of the equations from Lauwerier (1955) and Avdonin (1964) can be found in Spillette (1965). Figure 3.27 shows the comparison of the two equations considering linear flow. The only difference lies in the consideration of thermal conduction in the longitudinal direction of Equation 3.126.

The **coded functions** (MATLAB scripts) of Equations 3.122 and 3.126 are listed as *T_lau_linear.m* and *T_avd_linear.m*, respectively. As an example, Figure 3.28 shows the temperature response at 1 m downstream from the injection well. Consideration of thermal conduction in the longitudinal direction (Equation 3.126) results in a smooth temperature response.

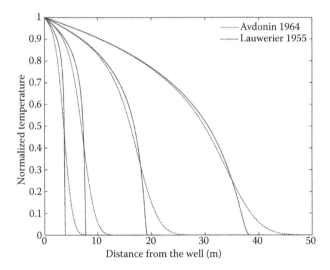

Figure 3.27 Normalized temperature $\bar{T} = (T - T_0)/(T_{inj} - T_0)$ as a function of distance for linear flow equations ($q_0 = 1 \times 10^{-5}$ m s^{-1}, $m = 10$ m, $\lambda_m = \lambda_{m2} = 2.5$ W m^{-1} K^{-1}). Temperature profiles shown at 10, 20, 50, and 100 days.

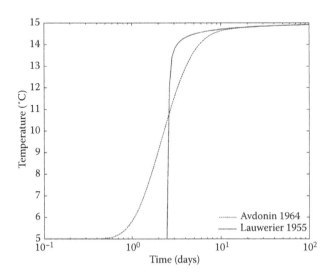

Figure 3.28 Temperature response over time at 1 m downstream from injection well ($q_0 = 1 \times 10^{-5}$ m s^{-1}, $m = 10$ m, $m\lambda_m = \lambda_{m2} = 2.5$ W m^{-1} K^{-1}). Equation 3.122: Lauwerier (1955); Equation 3.126: Avdonin (1964).

3.2.3 Radial flow, infinite disk source

Guimerà et al. (2007) modify the two-dimensional transient solute transport analytical solution after Gelhar and Collins (1971), which estimates contaminant distribution in porous media due to a fully penetrating injection well for zero flow conditions, used for calculating the temperature distribution due to an injection well of a groundwater heat pump system. The modified equation for horizontal conductive heat transport, from a continuous point source without groundwater flow after Guimerà et al. (2007), is given by

$$\overline{T} = \frac{T - T_0}{T_{inj} - T_0} = \frac{1}{2} \mathrm{erfc} \left[\frac{r^2 - R_{tw}^2}{2 \left[\frac{4\beta_L R_{tw}^3}{3} + \frac{\lambda_m R_{tw}^4}{C_m A_T} \right]^{1/2}} \right] \tag{3.128}$$

where R_{tw} represents the thermal radius of influence.

$$R_{tw} = \sqrt{2A_T t} \tag{3.129}$$

$$A_T = \frac{QC_w}{2\pi m C_m} \tag{3.130}$$

The previous analytical solution does not take axial effects into account. It assumes that there is no heat exchange with the upper and lower layers. For radial type flow, however, when there is a minor influence of the natural groundwater flow, axial effects might become important, especially for long-term simulations. Malofeev (1960) and Avdonin (1964) present analytical solutions with radial flow considering axial effects. Malofeev (1960) slightly modified the Lauwerier solution in order to apply it to radial flow. The dimensionless parameters shown in Equation 3.125 are changed as follows:

$$\xi = \frac{4\lambda_{m2}\pi r^2}{mC_w Q} \tag{3.131}$$

As in the linear flow, Avdonin (1964) adds thermal conduction in the horizontal direction, resulting in the following equation:

$$\overline{T} = \frac{1}{\Gamma(v)} \left[\frac{\overline{r}^2}{4\lambda_e \tau} \right]^v \int_0^1 \exp \left[-\frac{\overline{r}^2}{4\lambda_e \tau y} \right] \mathrm{erfc} \left(\frac{\sqrt{\tau/\lambda_e}\, \psi}{2a\sqrt{1-\psi^2}} \right) \frac{d\psi}{\psi^{v+1}} \tag{3.132}$$

$$v = \frac{QC_w}{4\pi m \lambda_m} \quad \text{and} \quad \overline{r} = 2r/m \tag{3.133}$$

Figure 3.29 shows the temperature response for the case considering radial flow. Avdonin (1964) and Malofeev (1960) account for heat transfer within the confining layers. Hence, there is higher dissipation of heat.

There are other approaches that consider a finite length of the overlying layer (Chen and Reddell 1983; Voigt and Haefner 1987) and surface heat exchange (Güven et al. 1983).

The **coded functions** (MATLAB script) of Equations 3.128, 3.131, and 3.132 are listed as *T_guimera.m*, *T_lau_radial.m*, and *T_avd_radial.m*, respectively. As an example, Figure 3.30 shows the temperature response at a radial distance of 1 m away from the injection well.

3.2.4 Natural background groundwater flow

To our knowledge, there is no exact analytical solution to simulate the temperature response of an aquifer considering an injection well and natural background groundwater flow. Therefore, we can only use in an approximation the closest exact analytical solutions for closed systems.

By considering the following energy relationship:

$$q_{tb} = \frac{QC_w \cdot (T_{inj} - T_0)}{m} \tag{3.134}$$

Equation 3.50, for instance, can be used, resulting in the following equation:

$$T(x,y,t) = T_0 + \frac{QC_w(T_{inj} - T_0)}{4\pi m C_m \sqrt{D_{t,L} D_{t,T}}} \exp\left[\frac{u_t x}{2D_{t,L}}\right]$$

$$\times \int_{\left(\frac{x^2}{4D_{t,L}} + \frac{y^2}{4D_{t,T}}\right)}^{\infty} \exp\left[-\psi - \left(\frac{x^2}{D_{t,L}} + \frac{y^2}{D_{t,T}}\right)\frac{u_t^2}{16D_{t,L}\psi}\right]\frac{d\psi}{\psi} \tag{3.135}$$

Making use of the **Hantush approximation**:

$$W(a_1, a_2) = \sqrt{\frac{\pi}{2a_2}} \exp(-a_2)\operatorname{erfc}\left[-\frac{a_2 - 2a_1}{2\sqrt{a_1}}\right] \tag{3.136}$$

we can express Equation 3.135 as follows:

$$T(x,y,t) = T_0 + \frac{QC_w \cdot (T_{inj} - T_0)}{4m C_m \sqrt{\pi D_{t,T} u_t r}} \exp\left[\frac{u_t \cdot (x - r)}{2D_{t,L}}\right]\operatorname{erfc}\left[\frac{r - u_t t}{2\sqrt{D_{t,L} t}}\right] \tag{3.137}$$

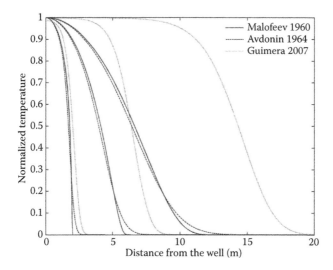

Figure 3.29 Normalized temperature $\bar{T} = (T - T_0)/(T_{inj} - T_0)$ as a function of distance for radial flow equations (Q = 9 m³ day⁻¹, m = 1 m, $\lambda_m = \lambda_{m2}$ = 2.5 W m⁻¹ K⁻¹). Temperature profiles shown at 1, 10, and 50 days.

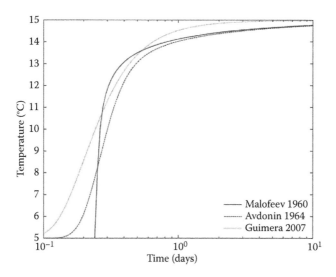

Figure 3.30 Temperature response over time at a radial distance of 1 m from injection well (Q = 9 m³ day⁻¹, m = 1 m, $\lambda_m = \lambda_{m2}$ = 2.5 W m⁻¹ K⁻¹). Equation 3.128: Guimerà et al. (2007); Equations 3.122 and 3.131: Malofeev (1960); Equation 3.132: Avdonin (1964).

where $r = \sqrt{x^2 + y^2 D_{t,L}/D_{t,T}}$. Assuming $\lambda_m = 0$, the above equation is reduced to the equation given by (Keim and Lang 2008)

$$T(x,y,t) = T_0 + \frac{QC_w \cdot (T_{inj} - T_0)}{4mC_m\sqrt{\pi\beta_T}\, u_t} \exp\left[\frac{x-r}{2\beta_L}\right]\frac{1}{\sqrt{r}}\,\mathrm{erfc}\left[\frac{r-u_t t}{2\sqrt{u_t\beta_L t}}\right] \qquad (3.138)$$

This analytical solution is currently used as a regulatory tool for installation, design, and management of open-loop systems in the Federal State of Baden-Württemberg, Germany (Baden-Württemberg 2009).

Using this approach, thermal plume lengths due to heat advection and conduction can be estimated. As explained before, however, this procedure does not consider the hydraulic influence of the injection well.

There are other approximations that account for groundwater flow. Kobus and Mehlhorn (1980), for instance, develop a two-dimensional analytical approximation for simulating transient axial heat conduction through the confining layers of a confined aquifer due to groundwater heat pump systems. The groundwater flow velocity is used for calculating the distance from the source location to a hypothetical line that represents the border between the natural groundwater streamlines and the streamlines diverted due to the local gradient around the injection well. Rauch (1992) developed an analytical formulation for heat transport in aquifers using the groundwater velocity parameter similar to Kobus and Mehlhorn. Additionally, based on a deviation angle of the groundwater flow direction and thermal dispersive transport analyses, the lateral temperature distribution in porous media due to heat conduction is predicted. The basic approach of the previous method is presented by Ingerle (1988) and is currently used in a regulatory guideline of the Austrian Association for Water and Waste Management (ÖWAV 2009).

REFERENCES

Abramowitz, M., Stegun, I.S. (1964). *Handbook of Mathematical Functions. With Formulas, Graphs and Mathematical Tables.* Dover Publ., (1980) Inc., New York, 1046p.

Avdonin, N.A. (1964). Some formulas for calculating the temperature field of a stratum subject to thermal injection. *Izvestijavyssichucebnychzavedenij/Neft' igaz* 7(3), 37–41.

Baden-Württemberg (2009). *Leitfaden zur Nutzung von Erdwärme mit Grundwasserwärmepumpen.* Federal State of Baden-Württemberg, Germany.

Bandos, T.V., Montero, Á., Fernádez, E., Santander, J.L.G., Isodro, J.M., Pérez, J., de Códoba, P.J.F., Urchueguía, J.F. (2009). Finite line-source model for borehole heat exchangers: Effect of vertical temperature variations. *Geothermics* 38, 263–270.

Beck, M., Bayer, P., de Paly, M., Hecht-Mendez, J., Zell, A. (2013). Geometric arrangement and operation mode adjustment in low enthalpy geothermal fields for heating. *Energy* 49,434–443.

Bear, J. (1979). *Hydraulics of Groundwater*. McGraw-Hill, New York.

Bernier, M.A., Pinel, P., Labit, R., Paillot, R. (2004). A multiple load aggregation algorithm for annual hourly simulations of GCHP systems. *HVAC&R Research* 10, 471–487.

BLOCON (2008). Earth Energy Designer 3.0. BLOCON, Lund, Sweden. Current Version 3.16.

Bredehoeft, J.D., Papadopulos, I.S. (1965). Rates of vertical groundwater movement estimated from the earth's thermal profile. *Water Resources Research* 1, 325–328.

Bundschuh, J. (1993). Modeling annual variations of spring and groundwater temperatures associated with shallow aquifer systems. *Journal of Hydrology* 142, 427–444.

Burger, A., Recordon, E., Bovet, D., Cotton, L., Saugy, B. (1984). *Thermique des nappes souterraines*. Presses Polytechniques Romandes, EPFL, Lausanne, Switzerland.

Carslaw, H.S., Jaeger, J.C. (1946). *Conduction of Heat in Solids*. 1st ed., Oxford University Press, Oxford, UK.

Carslaw, H.S., Jaeger, J.C. (1959). *Conduction of Heat in Solids*. 2nd ed., Oxford University Press, Oxford, UK.

Chaudhry, M.A., Zubair, S.M. (1994). Generalized incomplete gamma functions with applications. *Journal of Computational and Applied Mathematics* 55(1), 99–123.

Chen, C.-S., Reddell, D.L. (1983). Temperature distribution around a well during thermal injection and a graphical technique for evaluating aquifer thermal properties. *Water Resources Research* 19(2), 351–363.

Cui, P., Yang, H., Fang, Z. (2006). Heat transfer analysis of ground heat exchangers with inclined boreholes. *Applied Thermal Engineering* 26, 1169–1175.

DaCosta, J.A., Bennett, R.R. (1960). The pattern of flow in the vicinity of a recharging and discharging pair of wells in an aquifer having areal parallel flow. IAHS Publ. No. 52, pp. 524–536.

Diao, N., Li, Q., Fang, Z. (2004). Heat transfer in ground heat exchangers with groundwater advection. *International Journal of Thermal Sciences* 43(12), 1203–1211.

Domenico, P.A., Palciauskas, V.V. (1973). Theoretical analysis of forced convective heat transfer in regional ground-water flow. *Geological Society of America Bulletin* 84, 3803–3814.

Eskilson, P. (1987). Thermal analysis of heat extraction boreholes. Ph.D. Thesis Univ. of Lund, Lund, Sweden.

Eskilson, P., Claesson, J. (1988). Simulation model for thermally interacting heat extraction boreholes. *Numerical Heat Transfer* 13, 149–165.

Gelhar, L.W., Collins, M.A. (1971). General analysis of longitudinal dispersion in nonuniform flow. *Water Resources Research* 7(6), 1511–1521.

GLHEPRO (2007). GLHEPRO 4.0 for Windows User's guide. School of Mechanical and Aerospace Engineering Oklahoma State University. Distributed by the International Ground Source Heat Pump Association.

Gringarten, A.C., Sauty, J.P. (1975). A theoretical study of heat extraction from aquifers with uniform regional flow. *Journal of Geophysical Research* 80(35), 4956–4962.

Gröber, H., Erk, S., Grigull, U. (1955). *Grundgesetze der Wärmeübertragung.* U. Grigull (rev. ed.), Springer, Berlin, Germany.

Guimerà, J., Ortuño, F., Ruiz, E., Pérez-Paricio, A. (2007). Influence of ground-source heat pumps on groundwater. *Conference Proceedings*, European Geothermal Congress, Unterhaching, Germany, 30 May–1 June 2007.

Güven, O., Melville, J.G., Molz, F.J. (1983). An analysis of the effect of surface heat exchange on the thermal behavior of an idealized aquifer thermal energy storage system. *Water Resources Research* 19(3), 860–864.

Häfner, F., Sames, D., Voigt, H.D. (1992). *Wärme- und Stofftransport.* Springer Verlag, Berlin.

Ham, P.A.S., Schotting, R.J., Prommer, H., Davis, G.B. (2004). Effects of hydrodynamic dispersion on plume lengths for instantaneous bimolecular reactions. *Advances in Water Resources*, 27(8), 803–813.

Hecht-Méndez, J., de Paly, M., Beck, M., and Bayer, P. (2013). Optimization of energy extraction for vertical closed-loop geothermal systems considering groundwater flow. *Energy Conversion and Management*, 66, 1–10.

Hellström, G. (1991). Ground heat storage. Thermal analyses of duct storage systems. PhD Thesis, Lund University, Lund, Sweden.

Huber, A. (2008). *Software Manual Program EWS Version 4.0 Calculation of Borehole Heat Exchangers.* Huber Energietechnik AG, Zurich, Switzerland.

Incropera, F.P., Dewitt, D.P., Bergman, T.L., Lavine, A.S. (2007). *Introduction to Heat Transfer.* 5th ed., J. Wiley & Sons, Hoboken, N.J., USA.

Ingerle, K. (1988). Beitrag zur Berechnung der Abkühlung des Grundwasserkörpers durch Wärmepumpen. *Österreichische Wasserwirtschaft* 40, 11(12).

Ingersoll, L.R., Zobel, O.J., Ingersoll, A.C. (1948). *Heat Conduction; with Engineering, Geological and Other Applications.* 1st ed., McGraw-Hill, New York, 278p.

Ingersoll, L.R., Zobel, O.J., Ingersoll, A.C. (1954). *Heat Conduction; with Engineering, Geological and Other Applications.* Rev. and ext. ed., McGraw-Hill, New York, 325p.

Keim, B., Lang, U. (2008). *Thermische Nutzung von Grundwasser durch Wärmepumpen.* Umweltministerium Baden-Württemberg, State of Baden-Württemberg, Stuttgart, Germany.

Kobus, H., Mehlhorn, H. (1980). Beeinflussung von Grundwassertemperaturen durch Wärmepumpen. *gwf Wasser Abwasser* 121(6), 261–268.

Lamarche, L., Beauchamp, B. (2007). A new contribution to the finite line-source model for geothermal boreholes. *Energy and Buildings* 39(2), 188–198.

Lauwerier, H.A. (1955). The transport of heat in an oil layer caused by the injection of hot fluid. *Applied Scientific Research, Section A*, 5 (2–3) 145–150.

Lippmann, M.J., Tsang, C.F. (1980). Ground-water use for cooling: Associated aquifer temperature changes. *Ground Water* 18(5), 452–458.

Lu, N., Ge, S. (1996). Effect of horizontal heat and fluid flow on the vertical temperature distribution in a semiconfining layer. *Water Resources Research* 32(5), 1449–1453.

Malofeev, G.E. (1960). Calculation of the temperature distribution in a formation when pumping hot fluid into a well. *Neft'I Gaz* 3(7), 59–64.

Man, Y., Yang, H., Diao, N., Liu, J., Fang, Z. (2010). A new model and analytical solutions for borehole and pile ground heat exchangers. *International Journal of Heat and Mass Transfer* 53(13–14), 2593–2601.

Marcotte, D., Pasquier, P., Sheriff, F., Bernier, M. (2010). The importance of axial effects for borehole design of geothermal heat-pump systems. *Renewable Energy* 35(4), 763–770.

Mehlhorn, H., Spitz, K.-H., Kobus, H. (1981). Kurzschlussströmung zwischen Schluck- und Entnahmebrunnen – Kritischer Abstand und Rückströmrate. *Wasser und Boden* 33(4) 170–174.

Mercer, J.W., Faust, C.R., Miller, W.J., Pearson Jr., F.J. (1982). Review of simulation techniques for aquifers thermal energy storage (ATES). In VenTe Chow (Editor), Advances in Hydroscience, 13, 1–129. Academic Press, New York, USA.

Metzger, T., Didierjean, S., Maillet, D. (2004). Optimal experimental estimation of thermal dispersion coefficients in porous media. *International Journal of Heat and Mass Transfer* 47(14–16), 3341–2253.

Michopoulos, A., Kyriakis, N. (2009). Predicting the fluid temperature at the exit of the vertical ground heat exchangers. *Applied Energy* 86, 2065–2070.

Mikhailov, M.D., Özisik, M.N. (1984). *Unified Analysis and Solutions of Heat and Mass Diffusion.* J. Wiley & Sons, New York, USA.

Molina-Giraldo, N., Bayer, P., Blum, P. (2011a). Evaluating the influence of thermal dispersion on temperature plumes from geothermal systems using analytical solutions. *International Journal of Thermal Sciences* 50(7), 1223–1231.

Molina-Giraldo, N., Zhu, K., Blum, P., Bayer, P., Fang, Z. (2011b). A moving finite line source model to simulate borehole heat exchangers with groundwater advection. *International Journal of Thermal Sciences* 50(12), 2506–2513.

Nagano, K., Katsura, T., Takeda, S. (2006). Development of a design and performance prediction tool for ground source heat pump system. *Applied Thermal Engineering* 26, 1578–1592.

ÖWAV (2009). Thermische Nutzung des Grundwassers und des Untergrunds— Heizen und Kühlen (Thermal use of groundwater and underground— Heating and cooling). ÖWAV-Regelblatt 207, Österreichischer Wasser- und Abfallwirtschaftsverbands ÖWAV (Guideline 207 of the Austrian Water and Waste Management Association), Vienna, Austria.

Philippe, M., Bernier, M., Marchio, D. (2009). Validity ranges of three analytical solutions to heat transfer in the vicinity of single boreholes. *Geothermics* 38(4), 407–413.

Rauch, W. (1992). Ausbreitung von Temperaturanomalien im Grundwasser. Veröffentlichung Univ. Innsbruck No. 188.

Signorelli, S., Bassetti, S., Pahud, D., Kohl, T. (2007). Numerical evaluation of thermal response tests. *Geothermics* 36(2), 141–166.

Spillette, A.G. (1965). Heat transfer during hot injection into an oil reservoir. *Journal of Canadian Petroleum Technology* 4(4), 213–218.

Sutton, M.G., Nutter, D.W., Couvillion, R.J. (2003). A ground resistance for vertical bore heat exchangers with groundwater flow. *Journal of Energy Resources Technology ASME* 125(3), 183–189.

Tóth, J. (1963). A theoretical analysis of groundwater flow in small drainage basins. *Journal Geophysical Research* 68(16) 4795–4812.

Trüeb, E. (1976). Die Bedeutung des Grundwassers für die Versorgung der Schweiz. DVGW-Schriftenreihe Wasser 10.

Uffink, G.J.M. (1983). Dampening fluctuations in groundwater temperature by heat exchange between the aquifer and the adjacent layers. *Journal of Hydrology* 60, 311–328.

Voigt, H.D., Haefner, F. (1987). Heat transfer in aquifers with finite caprock thickness during a thermal injection process. *Water Resources Research* 23(12), 2286–2292.

Woods, K., Ortega, A. (2011). The thermal response of an infinite line of open-loop wells for ground coupled heat pump systems. *International Journal of Heat and Mass Transfer* 54, 5574–5587.

Yang, S.Y., Yeh, H.D. (2008). An analytical solution for modeling thermal energy transfer in a confined aquifer system. *Hydrogeology Journal* 16, 1507–1515.

Zeng, H.Y., Diao, N.R., Fang, Z.H. (2002). A finite line-source model for boreholes in geothermal heat exchangers. *Heat Transfer Asian Research* 31(7), 558–567.

Zubair, S., Chaudhry, M. (1996). Temperature solutions due to time-dependent moving-line heat sources. *Heat Mass Transfer* 32(3), 185–189.

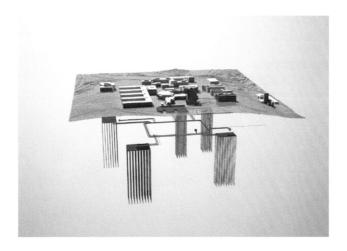

Figure 1.9 Heat storage project (schematic) at the Science City Campus (Hönggerberg) of ETH Zurich (Switzerland). (Courtesy of ETH Zurich, Abteilung Bau, 2011.)

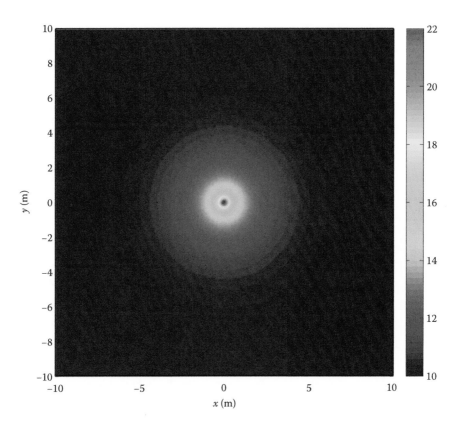

Figure 3.2 Temperature field for a single BHE with constant energy extraction after 90 days. ILS model (q_{tb} = J/H = 50 W m⁻¹, T_0 = 10°C, D_t = 9 × 10⁻⁷ m² s⁻¹).

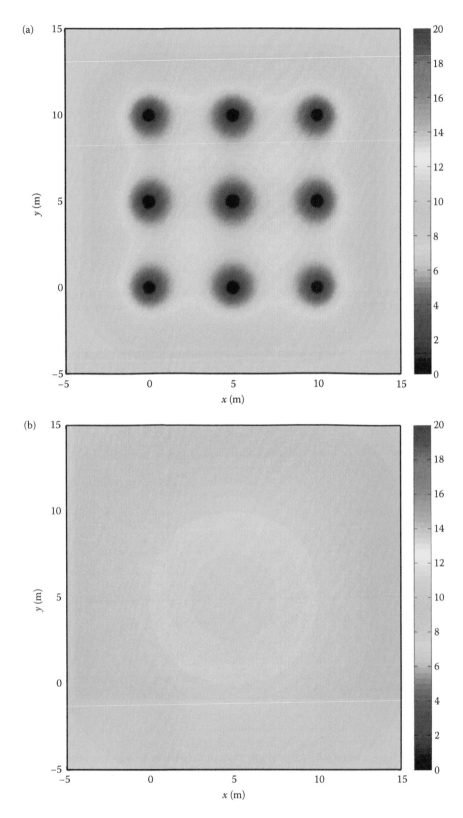

Figure 3.4 Heat exchanger group 3 × 3 calculated with ILS model with seasonal cosine heat input. (a) Map after 10.0 years; (b) 10.25 years; (c) 10.5 years; and (d) 10.75 years.

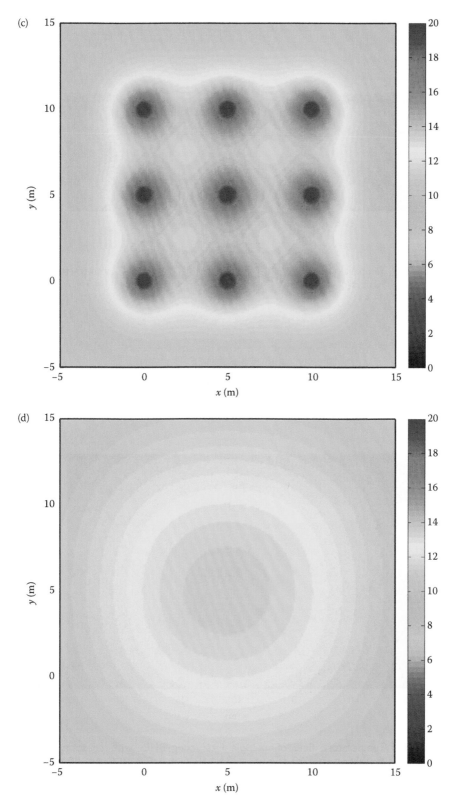

Figure 3.4 (Continued) Heat exchanger group 3 × 3 calculated with ILS model with seasonal cosine heat input. (a) Map after 10.0 years; (b) 10.25 years; (c) 10.5 years; and (d) 10.75 years.

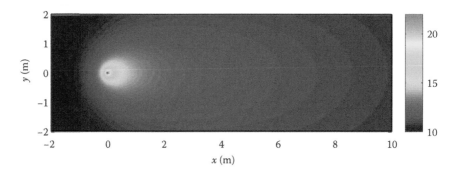

Figure 3.12 Temperature field for a single BHE with constant energy extraction after 90 days ($q_{tb} = J/H = 50 \, W \, m^{-1}$, $T_0 = 10°C$, $q = 1.0 \times 10^{-6} \, m \, s^{-1}$, $D_t = 9 \times 10^{-7} \, m^2 \, s^{-1}$).

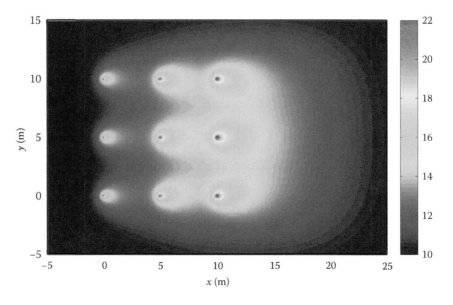

Figure 3.13 Temperature field of multiple interacting BHEs with constant energy extraction after 90 days ($q_{tb} = J/H = 50 \, W \, m^{-1}$, $T_0 = 10°C$, $q = 1.0 \times 10^{-6} \, m \, s^{-1}$, $D_t = 9 \times 10^{-7} \, m^2 \, s^{-1}$).

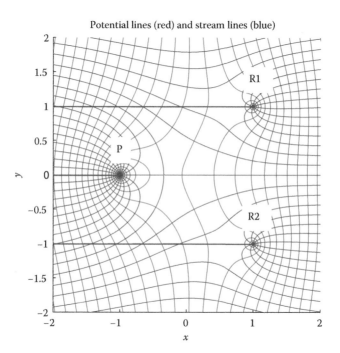

Figure 3.24 Scaled flow field with $\alpha = 0$ and scaled pumping rate; one pumping well and two injection wells. P: pumping well; R1, R2: recharge wells.

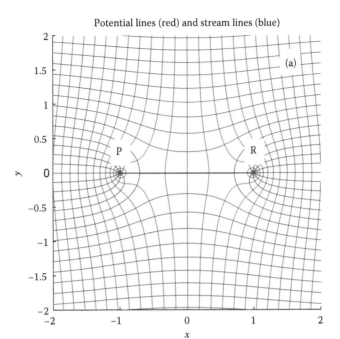

Potential lines (red) and stream lines (blue)

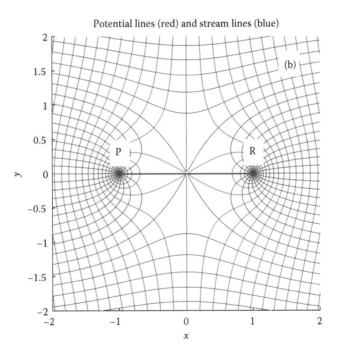

Potential lines (red) and stream lines (blue)

Figure 3.25 (a) Scaled flow field with $\alpha = 0$ and scaled pumping rate $\chi = 0.5$. P: pumping well; R: recharge well. (b) Scaled flow field with $\alpha = 0$ and $\chi = 1$. (c) Scaled flow field with $\alpha = 0$ and $\chi = 2$. (d) Scaled flow field with $\alpha = 90°$ and $\chi = 1$.

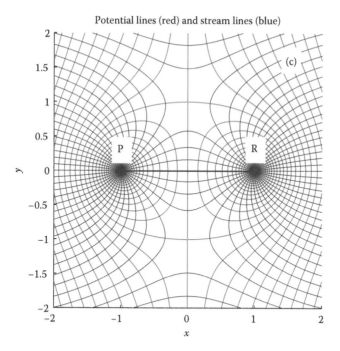

Potential lines (red) and stream lines (blue)

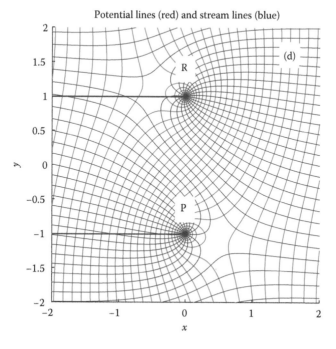

Potential lines (red) and stream lines (blue)

Figure 3.25 (Continued) (a) Scaled flow field with $\alpha = 0$ and scaled pumping rate $\chi = 0.5$. P: pumping well; R: recharge well. (b) Scaled flow field with $\alpha = 0$ and $\chi = 1$. (c) Scaled flow field with $\alpha = 0$ and $\chi = 2$. (d) Scaled flow field with $\alpha = 90°$ and $\chi = 1$.

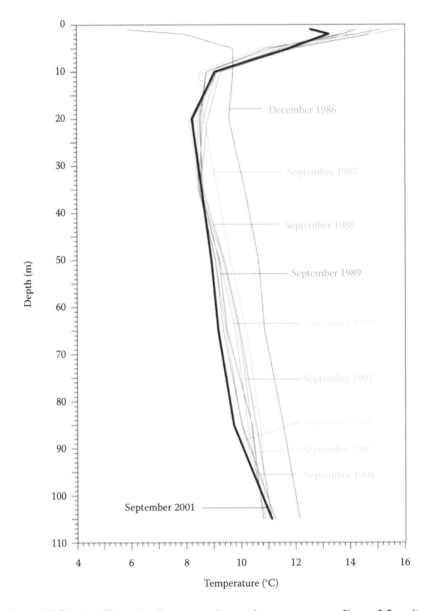

Figure 5.2 Elgg site (Switzerland): measured ground temperature profiles at 0.5 m distance from a 105 m deep operating BHE, repeatedly measured over 15 years. (From Rybach, L. and Eugster, W.J., *Geothermics* 39, 365–369, 2010.)

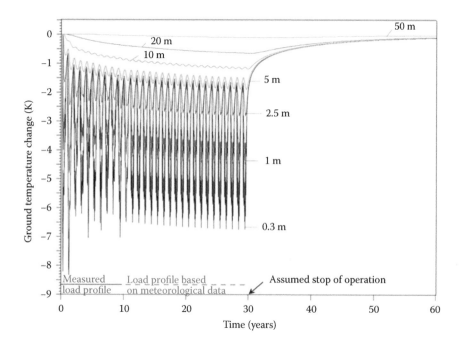

Figure 5.3 Elgg site (Switzerland): simulated ground temperature changes of a BHE relative to the undisturbed situation in December 1986 over 30 years of operation and 30 years of recovery. (From Rybach, L. and Eugster, W.J., *Geothermics*
39, 365–369, 2010.)

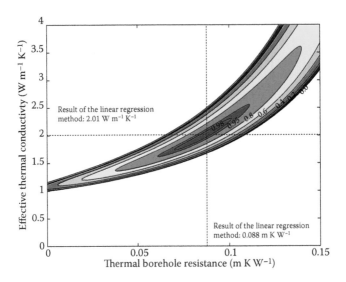

Figure 6.5 Example of the parameter estimation technique for the evaluation of the thermal borehole resistance and the effective thermal conductivities showing the results of the model efficiencies (EF values) according to Loague and Green (1991). The results of the linear regression method are also shown. (From Wagner 2010.)

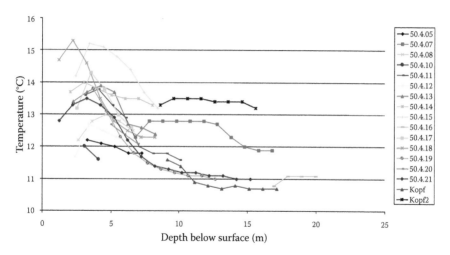

Figure 7.1 Altach study: temperature profiles in boreholes, measured on November 13, 2001. (Modified after Cathomen, N., Wärmetransport im Grundwasser, Auswirkungen von Wärmepumpen auf die Grundwassertemperatur am Beispiel der Gemeinde Altach im Vorarlberger Rheintal. Diploma thesis, ETH Zurich, Institute of Hydromechanics and Water Resources Management, 2002.)

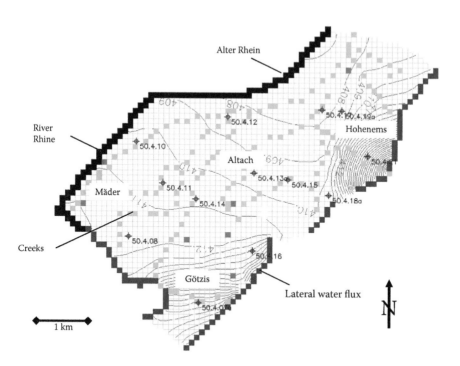

Figure 7.2 Altach study: two-dimensional flow model Altach (Austria) with head isolines (equidistance 1 m). Dark blue: river cells; blue: prescribed inflow cells; bright blue: creeks. (Modified after Cathomen, N., Wärmetransport im Grundwasser, Auswirkungen von Wärmepumpen auf die Grundwassertemperatur am Beispiel der Gemeinde Altach im Vorarlberger Rheintal. Diploma thesis, ETH Zurich, Institute of Hydromechanics and Water Resources Management, 2002.)

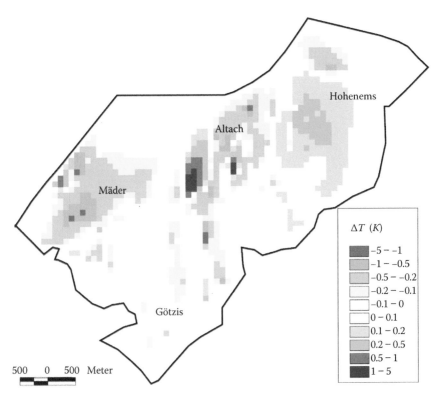

Figure 7.3 Altach study: two-dimensional heat transport model Altach (Austria) with temperature increase due to thermal use (groundwater heat pumps, heating by constructions). Dark blue: river cells; blue: prescribed inflow cells. (Modified after Cathomen, N., Wärmetransport im Grundwasser, Auswirkungen von Wärmepumpen auf die Grundwassertemperatur am Beispiel der Gemeinde Altach im Vorarlberger Rheintal. Diploma thesis, ETH Zurich, Institute of Hydromechanics and Water Resources Management, 2002.)

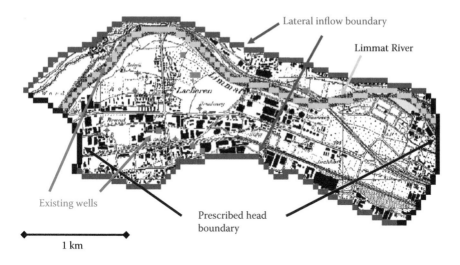

Figure 7.4 Case study Limmat Valley, subregion town of Schlieren: model domain with wells, and boundary conditions. (Modified after Müller, E., Ott, D., Thermische Nutzung des Grundwasserleiters Limmattal, Teilgebiet Hardhof-Schlieren [Thermal use of the Limmat Valley aquifer: Area Hardhof-Dietikon]. Report Master Project, ETH Zurich, Institute of Environmental Engineering, 2005.)

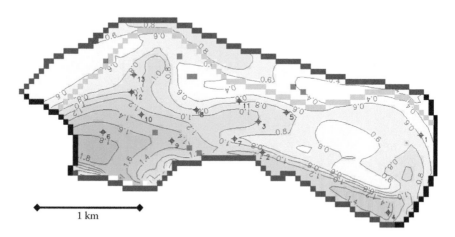

Figure 7.5 Case study Limmat Valley, subregion town of Schlieren: quasi-steady-state simulation of the temperature increase by warm basements. (Modified after Müller, E., Ott, D., Thermische Nutzung des Grundwasserleiters Limmattal, Teilgebiet Hardhof-Schlieren [Thermal use of the Limmat Valley aquifer: Area Hardhof-Dietikon]. Report Master Project, ETH Zurich, Institute of Environmental Engineering, 2005.)

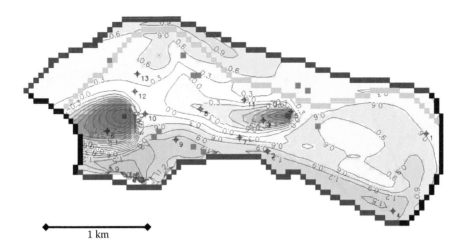

Figure 7.6 Case study Limmat Valley, subregion town of Schlieren: quasi-steady-state simulation of the temperature increase by warm basements, and, superimposed, two planned installations for the thermal use of groundwater. (Modified after Müller, E., Ott, D., Thermische Nutzung des Grundwasserleiters Limmattal, Teilgebiet Hardhof-Schlieren [Thermal use of the Limmat Valley aquifer: Area Hardhof-Dietikon]. Report Master Project, ETH Zurich, Institute of Environmental Engineering, 2005.)

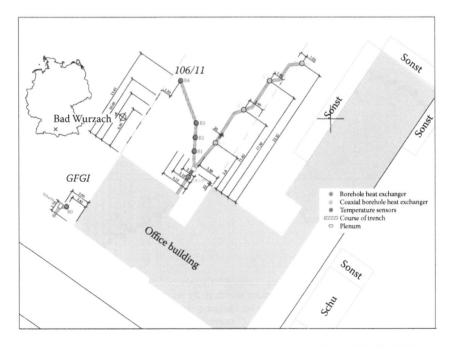

Figure 7.7 Bad Wurzach study: site map showing the locations of the BHE (EW1/09) and the five observation wells (B0, B1, B2, B3, and B4).

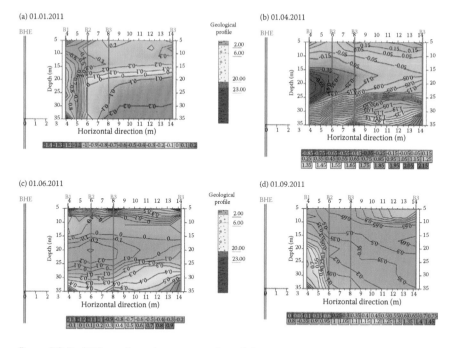

Figure 7.9 Bad Wurzach study: cross section of the temperature measurements in various depths in groundwater flow direction over a time period of one year.

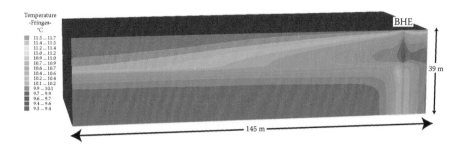

Figure 7.10 Bad Wurzach study: 3D numerical heat transport model of the BHE (EW1/09) showing the resulting temperature plume after 160.4 days.

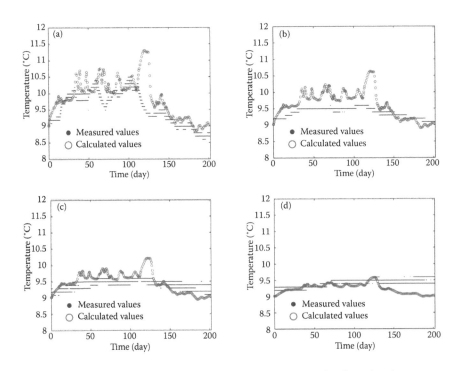

Figure 7.11 Bad Wurzach study: comparison between measured and simulated temperatures at the four observation wells (a) B1, (b) B2, (c) B3, and (d) B4 in 21 m depth in the vicinity of the BHE (EW1/09).

Chapter 4

Numerical solutions

The principal motivation for the **development and use of numerical models** for heat transport in subsurface environments was simulation of geothermal systems and heat storage in aquifers. In general, the development of numerical techniques was, to a large degree, anticipated by the development of models to simulate solute transport, starting in the 1970s. The **review** presented below concentrates on solutions of the advective and conductive heat transport problem in porous media, including both the saturated and unsaturated zones. Purely diffusive heat transport, a subject on which a vast number of contributions and models exist, is not considered here, since its numerical solution is formally identical to the solution of the diffusion equation (including the groundwater flow equation as a diffusion equation for pressure). Sometimes the solutions for pure heat conduction can be obtained from solutions for advection–dispersion–diffusion problems as special cases by setting the flow velocity equal to zero. The **heat transport problem** can be approximately solved in a **linearized form,** where temperature is influenced by flow, but flow is not influenced by temperature and density and hydraulic conductivity are assumed constant. This one-way coupling of flow and heat transport is the general assumption of authors. Alternatively, a **fully two-way coupled solution** is feasible where an iteration of the nonlinear system becomes necessary.

Mercer et al. (1982) presented a review on current simulation techniques for thermal energy storage in aquifers. Mercer et al. (1975) developed a transient two-dimensional model for the simulation of areal (horizontal) water flow and heat transport in a saturated aquifer, using the Galerkin finite element technique. Water viscosity and water density were taken as temperature dependent. They used the model to evaluate the hot-water geothermal system Wairakei (New Zealand), without taking into account phase-change processes. Their results were in general correspondence with the field data. Werner and Kley (1977) developed a three-dimensional finite difference model using cylindrical coordinates for the investigation of heat storage in aquifers. Radial flow velocity was assumed and dispersion effects were taken into account. They were able to approximately simulate a hydrothermal field experiment near Krefeld (Germany).

The **Lawrence Berkeley Laboratory** (Lippmann et al. 1977) developed the code CCC, which stands for conduction, convection, and consolidation, to simulate the coupled heat and momentum transport in one-, two-, and three-dimensional heterogeneous, anisotropic, nonisothermal porous media. Tsang et al. (1981) used this code to simulate the Auburn University field experiments (United States). They modeled two cycles of seasonal aquifer thermal energy storage (ATES). Simulated production temperatures and energy recovery factors agreed well with the field data.

Doughty et al. (1982) presented a dimensionless parameter approach to predict the thermal behavior of an ATES system. The analysis was restricted to radial flow in a horizontal aquifer confined by impermeable layers neglecting buoyancy effects. The heat transport equation was numerically integrated using an explicit finite difference approach.

Sauty et al. (1982) presented a theoretical study on the thermal behavior of a hot water storage system in an aquifer using a single well. They developed an axially symmetrical model, solved it applying a finite difference scheme, and checked it against analytical solutions. Buoyancy effects were neglected. The model was then used to evaluate the well temperature during production periods for symmetrical cycles (production volume and flow rate equal to injection volume and flow rate). They used both a fully implicit conductive scheme and an upstream explicit advective scheme. From the results, they deduced type curves for sets of dimensionless parameters.

Wiberg (1983) analyzed transient heat storage in an aquifer using the finite element method. His basic theory included nonlinear thermal physical properties and boundary conditions. Numerical simulations are shown for a purely conductive case with heat storage and a one-dimensional conductive–advective heat transport problem with a nonlinear decay term. Xue et al. (1990) used a three-dimensional alternating-direction implicit scheme to solve the heat transport equation. The flow was assumed as radial and heat transport included heat dispersion. The model was successfully used to investigate aquifer thermal heat storage in groundwater in China.

Merheb (1984) formulated a horizontal two-layer groundwater flow model and a corresponding heat transport model with heat exchange between layers and soil surface using a finite difference technique. The flow and the heat transport models were implemented in an uncoupled, sequential manner. He applied the model for the Strasbourg region (France).

Molson et al. (1992) formulated a three-dimensional finite element model for simulating coupled density-dependent groundwater flow and heat transport in aquifers. The heat transport solution is based on a finite element time integration, which generates a symmetrical coefficient matrix. The thermal transport model was successfully checked against the results of the Borden (Canada) thermal injection field experiment. Dwyer and Eckstein

(1987) formulated a two-dimensional, horizontal, Galerkin finite element model for a feasibility study of ATES coupled with a heat pump. The flow and the heat transport models were applied in an uncoupled, sequential manner. In heat transport, advection and mechanical dispersion were taken into account.

Sun and Carrington (1995) developed a so-called implicit correction scheme for advection-dominated heat transfer in porous media with strong temperature gradient. The scheme allows relatively coarse grid size for the numerical discretization. Chevalier and Banton (1999) used the random walk method to study heat transfer problems in porous media with a radial flow field. Buoyancy effects were disregarded and none of the physical properties were dependent on the temperature. They checked the model against analytical and numerical solutions. Kohl and Hopkirk (1995) presented the simulation code FRACTure for forced water flow in fractured rocks. Hydrodynamics were coupled to rock mechanics but not to heat transport. They applied the code to hot dry rock sites. Signorelli et al. (2007) used this code for a numerical evaluation of thermal response tests. Hecht-Méndez et al. (2010) used MT3DMS (Zheng and Wang 1999) to simulate heat transport in closed geothermal systems, assuming that buoyancy effects and temperature dependency of water viscosity are negligible. They compared their results with those of analytical solutions and numerical solutions using SEAWAT (Langevin et al. 2008) and found good agreement.

Diersch and Kolditz (1998) analyzed double-diffusion and buoyancy driven free-convection processes using the code FEFLOW (Diersch 1996).

Chiasson et al. (2000) numerically investigated the effects of groundwater flow on closed-loop ground-source heat pump systems and postulated that heat advection can significantly enhance heat transfer from and to borehole heat exchangers (BHEs). Similar conclusions were drawn by Fan et al. (2007). An initial assessment of the importance of advection can be obtained by an examination of the thermal Peclet number (Chiasson et al. 2000).

Ferguson (2007) examined the effect of heterogeneities on heat transport by stochastic modeling, including dispersion effects using geostatistics of aquifers. His results indicate that there is considerable uncertainty in the distribution of heat associated with the injection of warm water into an aquifer. Advective–conductive heat transport models were created using METRA, which is a submodule of the code MULTIFLOW (Painter and Seth 2003). METRA is an integrated finite difference code capable of simulating variable-density fluid flow and heat flow. Hidalgo et al. (2009) performed a Monte Carlo analysis of steady-state advective–conductive heat transfer in heterogeneous aquifers using a finite element code.

Graf and Therrien (2007) formulated a model for coupled fluid flow, heat, and single-species reactive mass transport with variable fluid density and viscosity in fractured porous media. The effects were incorporated in the code HydroGeoSphere. Brookfield et al. (2009) performed a numerical study on thermal transport modeling in a fully integrated surface–subsurface framework using this code.

Engeler et al. (2011) investigated heat transport in an aquifer with strong river–aquifer interaction. They used the code SPRING (delta-h 2012), which uses three-dimensional finite elements and allows temperature dependence of the flow parameters. They showed that better agreement with measured temperature data is obtained if the temperature dependence of the leakage coefficient (via the temperature dependence of viscosity) is taken into account in the modeling.

Several authors have also investigated heat transport in unsaturated porous media. Sophocleous (1979) formulated an implicit vertical finite difference model for the analysis of coupled nonlinear water and heat transport under saturated and unsaturated conditions. He used and extended the Philip and de Vries (1957) formulation of coupled nonisothermal flow of water, vapor, and heat (Parlange et al. 1998). Yeh and Luxmore (1983) presented a multidimensional model for moisture and heat transport in unsaturated porous media using the so-called integrated compartment method, which is an extension of the integrated finite difference method. Again, the Philip and de Vries (1957) nonisothermal equations were used for simultaneous moisture and heat transport. Sidiropoulos and Tzimopoulos (1983) performed a sensitivity analysis of coupled water and heat transfer in porous media. For their case, they found that phase-change effects could be neglected. Birkholzer and Tsang (2000) used the code TOUGH2 (Wu et al. 1996) for the modeling of the coupled thermohydraulic processes in a large-scale underground heater test in partially saturated fractured tuff.

Al-Khoury (2012), Al-Khoury and Bonnier (2006), and Al-Khoury et al. (2005, 2010) presented computationally efficient finite element tools for the analysis of three-dimensional steady-state and transient heat flow in geothermal systems. They assumed that temperature has no influence on groundwater flow. They formulated one-dimensional heat pipe finite elements, which are capable of simulating pseudo-three-dimensional heat flow in a vertical BHE consisting of pipe-in, pipe-out, and grout material. Three-dimensional finite elements for saturated aquifers were formulated, which can be in contact with heat pipe finite elements. Their method was extended by Bauer et al. (2011) and Diersch et al. (2011a,b) and incorporated in the software FEFLOW (DHI-WASY 2010).

Glück (2011) developed engineering software for the numerical simulation of underground heat exchangers. Steady-state and transient axially symmetrical temperature fields due to heat conduction are calculated using the finite volume method. Thermal processes within the BHE with inflow and outflow tubes (single- and double-U-tube configuration) embedded in grouting material are restricted to quasi-steady-state conditions and are evaluated using the concept of heat transfer coefficients. By evaluating an effective radius of a single device, regular fields of BHEs are approximated.

Lazzari et al. (2010) investigated the long-term performance of BHE fields with negligible groundwater movement by finite elements using the software package COMSOL. Lee and Lam (2008) performed computer simulations for BHE systems using the finite difference approach. Outside the borehole, heat transport is restricted to heat conduction. Inside the borehole, flow in the tubes is incorporated. Fujimitsu et al. (2010) numerically evaluated the environmental impact caused by a ground-coupled heat pump system using the FEFLOW software. Park et al. (2012) investigated the heat transfer of helical BHEs experimentally, analytically, and numerically. Park et al. (2013) numerically modeled precast, high strength concrete energy piles. Jalaluddin and Miyara (2012) numerically investigated the performance of several types of vertical BHEs in continuous and discontinuous operation modes with the software FLUENT.

Laloui et al. (2006) formulated a coupled displacement, pore water pressure, temperature finite element model for heat exchanger piles. The model was able to reproduce in situ experimental observations.

Deng et al. (2005) suggested and tested a simplified numerical model for the simulation of standing column well ground heat exchangers. Woods and Ortega (2011) numerically investigated the thermal response of a line of standing column wells and compared these results with analytical models.

Kim et al. (2010) numerically investigated the performance of ATES systems in confined aquifers (open systems). They formulated a three-dimensional aquifer flow and heat transport model with finite elements, assuming constant water density and viscosity, using COMSOL. They concluded that the thermal interference of an ATES system (affecting primarily the system performance) depends on the distance between the two boreholes, the hydraulic conductivity of the aquifer, and the production/injection rate. The thermal interaction of pumping and injecting well groups with absent regional groundwater flow was numerically investigated by Gao et al. (2013). They assumed that material properties do not depend on temperature.

4.1 TWO-DIMENSIONAL HORIZONTAL NUMERICAL SOLUTIONS

Two-dimensional numerical solutions are mainly discussed here with respect to their application concerning **open thermal systems**, that is, systems with water abstraction and reinfiltration after temperature increase or decrease by ΔT. The **vertically integrated, two-dimensional heat transport equation for shallow regional aquifers** focuses on the saturated part, as recalled here:

$$\frac{\partial T}{\partial t} = \nabla \cdot (\mathbf{D}_t \nabla T) - \frac{C_w}{C_m} \nabla \cdot (\mathbf{q}T) + \frac{P_t}{mC_m} + \frac{J_{vert,bot}}{mC_m} + \frac{J_{vert,top}}{mC_m} \qquad (4.1)$$

The variable $T(\mathbf{x}, t)$ is the mean temperature in a vertical profile at location $\mathbf{x}$ and time t. Vertical heat flow from below (geothermal heat flux), $J_{vert,bot}(\mathbf{x}, t)$, and vertical advective and diffusive transport from soil surface to aquifer, $J_{vert,top}(\mathbf{x}, t)$, are taken into account through source/sink terms. While the former can be expressed and inserted directly, the latter can be treated in a different manner. In the application of the two-dimensional heat transport equation, it is often assumed that the coupling of flow and transport via density can be neglected.

One possibility consists of inserting the **linear approximation for the vertical heat flux** $J_{vert,top}(\mathbf{x}, t)$ into the heat transport equation:

$$\frac{\partial T}{\partial t} = \nabla \cdot (\mathbf{D}_t \nabla T) - \frac{C_w}{C_m} \nabla \cdot (\mathbf{q}T) + \frac{P_t}{mC_m} + \frac{J_{vert,bot}}{mC_m} + \frac{\lambda_{vert} \cdot (T_{surface} - T)}{mC_m \cdot (f + m/2)}$$

$$+ \frac{NC_w T_{surface}}{mC_m} \qquad (4.2)$$

Equation 4.2 is expressed here as the transient balance equation. However, we have to keep in mind that the transient behavior is not fully considered, or only in a rudimentary way, for the linear flux terms from soil surface to groundwater. This disadvantage is avoided for **long-term, steady-state flow and heat transport** according to

$$\nabla \cdot (\mathbf{D}_t \nabla T) - \frac{C_w}{C_m} \nabla \cdot (\mathbf{q}T) + \frac{P_t}{mC_m} + \frac{J_{vert,bot}}{mC_m} + \frac{\lambda_{vert} \cdot (T_{surface} - T)}{mC_m \cdot (f + m/2)}$$

$$+ \frac{NC_w \cdot (T_{surface} - T_0)}{mC_m} = 0 \qquad (4.3)$$

Furthermore, if we adopt the considerations pointed out in Chapter 2.1.3.1 to formulate the equation in terms of **temperature differences** $\Delta T(\mathbf{x}) = T(\mathbf{x}) - T_{surface}(\mathbf{x})$ and to choose $T_0 = T_{surface}$ assuming **constant surface temperature**, Equation 4.3 reduces to

$$\nabla \cdot (\mathbf{D}_t \nabla T) - \frac{C_w}{C_m} \nabla \cdot (\mathbf{q}T) + \frac{P_t}{mC_m} + \frac{J_{vert,bot}}{mC_m} + \frac{\lambda_{vert} \cdot (T_{surface} - T)}{mC_m \cdot (f + m/2)} = 0 \qquad (4.4)$$

The equation formally corresponds to the steady-state solute transport equation with first-order decay.

A main contribution to the heat production term P_t is the heat flux from **large warm basements of constructions.** The vertical contribution can be approximately expressed according to Equation 2.125 by

$$
J_{\text{cond\_basement}} = A_{\text{basement}} \lambda_{\text{vert}} \frac{T_{\text{basement}} - T_{\text{gw}}}{\left(f + \dfrac{m}{2} \right)_{\text{basement}}} = A_{\text{basement}} \lambda_{\text{vert}} \frac{(T_{\text{basement}} - T_0) - (T_{\text{gw}} - T_0)}{\left(f + \dfrac{m}{2} \right)_{\text{basement}}}
$$

$$
= A_{\text{basement}} \lambda_{\text{vert}} \frac{(T_{\text{basement}} - T_0) - T_{\text{rel}}}{\left(f + \dfrac{m}{2} \right)_{\text{basement}}} \tag{4.5}
$$

Therefore, the heat flux consists of a first-order part in $T_{\text{rel}} = T_{\text{gw}} - T_0$ and a constant heat flux. The latter represents a heat-loading rate.

Another interesting possibility consists of **coupling the two-dimensional heat transport equation for the (saturated) regional aquifer with the one-dimensional vertical heat-flow between soil surface and aquifer.** In water flow problems, this coupling has been realized in HYDRUS-1D-MODFLOW (Seo et al. 2007). It can be achieved, in principle, by adding vertical columns on top of each finite difference cell or finite element. In practice, however, areas of identical parameters and conditions are defined within the solution domain D in order to reduce the computational effort.

In the case of **confined aquifers,** the **vertical heat flux** $J_{\text{vert,top}}(\mathbf{x}, t)$ at the top of the aquifer is described by the heat transport equation in solids, according to Equation 2.101, here without production term:

$$
\frac{\partial T}{\partial t} = \frac{\partial}{\partial z} \left[D_t \frac{\partial T}{\partial z} \right] \tag{4.6}
$$

Coupling of the two models requires the continuity of temperature and the continuity of the heat flux.

For **unconfined aquifers, the vertical heat balance equation,** that is,

$$
\frac{\partial T}{\partial t} = \frac{\partial}{\partial z} \left[(D_{t,\text{diff}} + D_{t,\text{disp,L}}) \frac{\partial}{\partial z} T \right] - \frac{C_w}{C_m} \frac{\partial}{\partial z} (q_z T) \tag{4.7}
$$

has to be coupled with the water flow equation, that is, the **vertical Richards equation:**

$$
\phi \frac{\partial S_w}{\partial t} = \frac{\partial}{\partial z} \left(K_w(S_w) \frac{\partial}{\partial z} \left(z + \frac{p_w}{\rho_w g} \right) \right) \tag{4.8}
$$

Coupling of the models now requires the continuity of temperature and water, as well as heat fluxes at the interface between the two regions. With this procedure, the computational effort can be reduced considerably compared to a three-dimensional saturated–unsaturated model. In principle, the coupling with vertical soil columns can also be undertaken in connection with a multilayer flow and heat transport model.

4.1.1 Analogy with solute transport models

Obviously, one may **utilize any existing solute transport code** for heat transport as well, if one establishes the analogy between both models in a consistent way. The principle is shown using the example of Equation 4.4. A comparison with the **steady-state solute transport equation**, that is,

$$\nabla \cdot (\mathbf{D}_h \nabla c) - \frac{1}{\phi} \nabla \cdot (qc) + \frac{P_c}{m\phi} - \lambda_c c = 0 \qquad (4.9)$$

yields the following correspondence between the parameter

Solute transport	Heat transport
Solute concentration $c \geq 0$	Temperature $T_{rel} \geq 0$
Hydrodynamic dispersion tensor $\mathbf{D}_h$	Thermal dispersion tensor $\mathbf{D}_h$
Molecular diffusion coefficient D_{mol}	Thermal diffusion coefficient D_t
Macrodispersivities α_L, α_T	Macrodispersivities β_L, β_T
Inverse porosity $1/\phi_c$	Thermal capacity ratio C_w/C_m
Source/sink term $P_c/(\phi m)$	Thermal source/sink $P_t/(C_m\, m)$
Decay coefficient λ_c	Thermal flux coefficient $\lambda_{vert}/(C_m m(f + m/2))$

Therefore, in order to simulate heat transport using a solute transport code, the parameters have to be defined as follows. In the solute transport code, **equivalent porosity** is

$$\phi_c = \frac{C_m}{C_w} \qquad (4.10)$$

The **equivalent solute mass production** term is

$$P_c = \frac{P_t \phi}{C_m} \qquad (4.11)$$

and the **equivalent decay coefficient** is

$$\lambda_c = \frac{\lambda_{vert}}{C_m m \cdot \left(f + \dfrac{m}{2} \right)} \tag{4.12}$$

Consequently, the **heat injection** rate $J_t = QC_w\Delta T$ translates into the analogous solute mass flux $J_c = Q\Delta c$. Note that since solute concentrations have to be positive, with $c \geq 0$, the relative temperature has to be positive as well, with $T_{rel} \geq 0$.

The **assumption of constant mean surface temperature** represents a simplification of the complex processes at the soil surface (see Chapter 1). In fact, it cannot be excluded that an increase or decrease in groundwater temperature could also affect the temperature at the soil surface. Such a situation could arise, for example, for aquifers with very small depth to groundwater.

We are not aware that codes exist that solve Equation 4.4 or its transient form. However, several codes exist that solve the **solute transport equation with first-order decay**, like MT3D (Zheng 1990), MT3DMS (Zheng and Wang 1999), HydroGeoSphere (Graf and Therrien 2007; Raymond et al. 2011), FEFLOW (DHI-WASY 2010), SPRING (delta-h 2012), and others. They can be used as analogs in order to solve thermal processes in two-dimensional aquifers. In any case, a simulation of thermal or solute transport requires the simulation of the water flow field beforehand.

4.1.2 Analysis of steady-state open system in rectangular aquifer

The procedure for simulating heat transport using a solute transport code is illustrated for the **example** of a simple rectangular confined aquifer of size 2000 m × 1000 m, of thickness $m = 12$ m, and with hydraulic conductivity $K_w = 0.002$ m s^{-1} (Figure 4.1). No areal recharge occurs. Piezometric head is

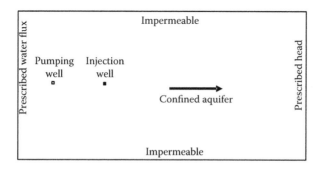

Figure 4.1 Illustrative example of two-dimensional aquifer: situation and hydraulic boundary conditions.

specified at the western boundary with h_w = 12 m. At the eastern boundary, the water inflow rate is specified with a total flow rate of Q_{inflow} = 0.036 m³ s⁻¹. Without wells, this corresponds to a flow gradient of 0.0015. Impermeable boundaries are present at the northern and southern boundaries. Thermal use is planned for an open system with extraction well, heat pump, and infiltration well. The pumping rate is Q_w = 1000 m³ day⁻¹. Since thermal use occurs in the winter season only, Q_w corresponds to a yearly average value. The distance between extraction and infiltration well is chosen in order to avoid hydraulic short-circuiting for the actual pumping rate in the winter season. The steady-state flow field is shown in Figure 4.2 with results from particle tracking indicating flow lines. The equivalent porosity value is ϕ_c = 0.571, using C_m = 2.4 × 10⁶ W m⁻³ K⁻¹. Markers on the particle tracks are shown in yearly intervals based on the velocity u_t of the thermal front.

The longitudinal thermal macrodispersivity is chosen as β_L = 10 m and the transversal macrodispersivity as β_T = 1 m. The size of the finite difference cells is 10 m. Boundary conditions are prescribed temperature T_{rel} = 0 at the eastern boundary, impermeable northern and southern boundaries, and the transmission boundary type (Chapter 2.1.2.5) for the western boundary. The latter is approximated by setting longitudinal dispersivity to zero along the outflow boundary. Water is pumped at the abstraction well. A heat pump lowers the temperature by 3 K. This is realized by setting a positive source concentration of c_{well} = 3.0. The equivalent decay coefficient is calculated using Equation 4.12, with λ_{vert} = 2.0 W m⁻¹ K⁻¹ and f = 6 m, which yields λ_c = 5.787 × 10⁻⁹ s⁻¹. The transport module was run for 20 years in order to approach steady-state thermal conditions. The thermal plume after 20 years of infiltration of cold water is shown in Figure 4.3. Note that this type of steady-state (time-averaged) analysis does not take into account the dynamics caused by seasonally varying pumping and infiltration rates. The

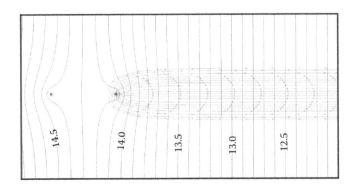

Figure 4.2 Illustrative example of two-dimensional aquifer: steady-state flow field with step in piezometric head Δh_w = 0.1 m. Particle tracks starting at infiltration well. Particle markers are introduced with increment Δt = 1 year (with respect to thermal velocity).

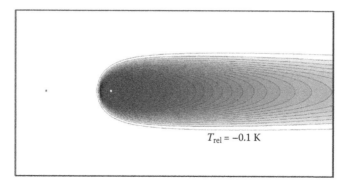

$T_{\text{rel}} = -0.1$ K

Figure 4.3 Illustrative example of two-dimensional aquifer: calculated thermal plume due to infiltration of cold water with $\Delta T = 3$ K, after 20 years. Temperature increment is 0.1 K.

latter would cause more lateral spreading of the thermal plume during the cold season and, therefore, also in the long run. Furthermore, it is important to realize that macrodispersion effects are usually overestimated close to the source, that is, the infiltration well. This has to do with the scale-effect of macrodispersivity as described in Chapter 2.1.2.3.

The example was calculated using standard MODFLOW-96 (USGS 2012) and MT3DMS (Zheng and Wang 1999), which are both incorporated in the code PMWIN (Chiang and Kinzelbach 2005). Transport was simulated using an upstream finite difference scheme with discretization $\Delta x = \Delta y = 10$ m and a time step corresponding to a Courant number of 0.75 (see Equation 4.32). The grid Peclet number (see Equation 4.30) was 2.

4.1.2.1 Scaled solution for open system in rectangular aquifer

For the simple rectangular layout of the illustrative example, the procedure can be extended by **scaling**. By introducing a **length scale** L, any length can be scaled such as $x' = x/L$ or $h'_w = h_w/L$. The **steady-state scaled form of the two-dimensional flow equation** 2.38 is

$$\nabla'^2 h'_w + \frac{NL}{K_w m} = 0 \qquad (4.13)$$

where ∇'^2 is the Laplace operator, in scaled form. We may choose the length scale as $L = Q/(mq_0)$ using the pumping and infiltration rate Q and the specific inflow rate q_0. Note that in this case, L is the recharge width of the flow toward the well.

Using the specified inflow rate q_0 to scale the water fluxes, and introducing a temperature scale Θ, the **scaled steady-state heat transport equation** without P_t and $J_{\text{vert,bot}}$ is

$$\frac{\beta_L}{L}\nabla'\cdot(\mathbf{D}'_t\nabla T'_{rel})-\nabla'\cdot(\mathbf{q}'T'_{rel})+\frac{\lambda_{vert}T'_{rel}Q}{(f+m/2)m^2 q_0^2 C_w}=0 \qquad (4.14)$$

assuming that macrodispersion is dominant compared to thermal diffusion. The temperature scale may be chosen as $\Theta = \Delta T$. The **decay term** can be written as $\lambda'_{vert}T'_{rel}$ with the dimensionless decay coefficient λ'_{vert}:

$$\lambda'_{vert}=\frac{\lambda_{vert}Q}{(f+m/2)m^2 q_0^2 C_w} \qquad (4.15)$$

A **time scale** τ can be obtained by setting $\tau = L/u_t = LC_m/(C_w q_0)$.

The **scaled temperature field is evaluated numerically** and analyzed for the scaled temperature profile along the streamline through the infiltration well. For $\lambda'_{vert} = 0.01$, 0.1, and 1, the dimensionless longitudinal dispersivity $\beta'_L = \beta_L/L = 0.02$ (which represents a small value), and the dispersivity ratio $\beta'_T/\beta'_L = 0.1$, the scaled temperature profile is shown in Figure 4.4. For $\beta_L = 0.2$ (representing a medium value) and again $\beta'_T/\beta'_L = 0.1$, it is shown in Figure 4.5. Obviously, there is quite some impact of dispersion on the shape of the plume. We can state that the shape is mainly governed by decay. Again, one has to be aware that dispersion effects are overestimated close to the infiltration well.

The numerical scaling analysis was again performed using standard MODFLOW-96 (USGS 2012) and MT3DMS (Zheng and Wang 1999), which are both incorporated in the code PMWIN (Chiang and Kinzelbach 2005).

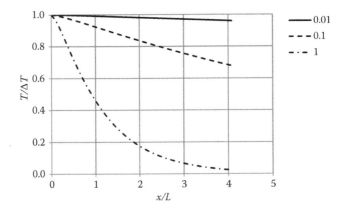

Figure 4.4 Scaled steady-state temperature profile downstream of the infiltration well in simple two-dimensional aquifer for dimensionless decay coefficients $\lambda'_{vert} = 0.01$, 0.1, 1 and $\beta'_L = 0.2$, $\beta'_T = 0.02$.

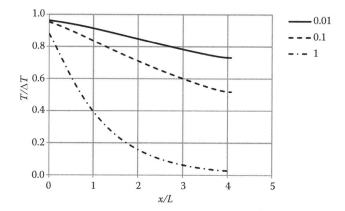

Figure 4.5 Scaled steady-state temperature profile downstream of the infiltration well in simple two-dimensional aquifer for dimensionless decay coefficients $\lambda'_{vert} = 0.01$, 0.1, 1 and $\beta'_L = 0.2$, $\beta'_T = 0.02$.

Transport was simulated using an upstream finite difference scheme with discretization $\Delta x' = \Delta y' = 0.02$ and a time step corresponding to a Courant number of 0.75 (see Equation 4.32). Results are shown for the dimensionless time $t' = 10$.

4.2 MULTIDIMENSIONAL NUMERICAL SOLUTIONS

Solution techniques are briefly discussed with respect to the **solution of the heat transport equation** 2.92 including heat advection:

$$\frac{\partial T}{\partial t} = \nabla \cdot [D_t \nabla T] - \frac{C_w}{C_m} \nabla \cdot (qT) + \frac{P_t}{C_m} \tag{4.16}$$

Frequently used **numerical solution methods** can be classified as follows:

- Finite difference method
- Finite element method
- Finite volume method
- Method of characteristics
- Random walk method

Note that in each group, various submethods and techniques have been formulated. In the following, we concentrate on basic ideas of the methods.

4.2.1 Principles of the finite difference method for heat transport

For the presentation of the principles, we prefer to start from the **energy balance equation**:

$$C_m \frac{\partial T}{\partial t} = \nabla \cdot \left[(\lambda_m + C_w D_{t,disp}) \nabla T \right] - C_w \nabla \cdot (\mathbf{q} T) + P_t \qquad (4.17)$$

The basic procedure of the **finite difference method** can be stated as follows: The solution domain D is **discretized into prismatic cells**. In the case of one- and two-dimensional problems, the shape reduces to linear and rectangular cells. Numbering of the cells is chosen according to three-dimensional matrices with layer, row, and column indices i, j, k. The cell size does not need to be constant. However, neighboring cells should still have similar size (change less than a factor of 2) for numerical reasons.

For each cell, the mean temperature $T_{ijk}(t)$ is either unknown (and to be calculated) or known (prescribed temperature). This temperature corresponds to the average value within the cells. The temperature $T_{ijk}(t)$ is assigned to the cell center.

For each cell (i, j, k), the **physical heat balance** is expressed over all surfaces of the cell using unknown temperatures $T_{ijk}(t + \Delta t)$ within the cells, given known temperatures in each cell, where Δt is the time step. The advective, diffusive, and dispersive heat fluxes are expressed by linear approximations, using the cell center temperatures from the neighboring cells. The **rate of change of the energy within the cell** is expressed using the time step Δt and the temperature difference $T_{ijk}(t + \Delta t) - T_{ijk}(t)$.

Boundary conditions (like prescribed temperature or prescribed heat flux) are directly considered in the corresponding balance equations for the cells. The resulting **equation system** is linear in the unknown temperatures at time ($t + \Delta t$). After obtaining the solution, the new temperatures $T_{ijk}(t + \Delta t)$ in the cells are the **initial conditions for a new time step**.

The **advective heat flux** in x-direction into the cell (i, j, k) (Figure 4.6) can be expressed as follows:

$$J_{x,i,j,k,adv}(t^*) = q_{x,i-\frac{1}{2},j,k}(t^*) C_w T_{i-\frac{1}{2},j,k}(t^*) \Delta y \Delta z$$

$$- q_{x,i+\frac{1}{2},j,k}(t^*) C_w T_{i+\frac{1}{2},j,k}(t^*) \Delta y \Delta z \qquad (4.18)$$

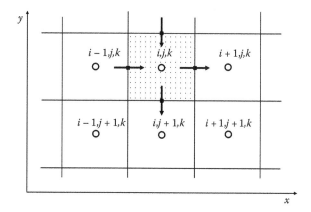

Figure 4.6 Finite difference grid with cell *i,j,k*.

The time t^*, at which the fluxes are determined, has to be specified. The index $i - 1/2$ and $i + 1/2$, respectively, denote the values at the interfaces between the cells. This formulation presumes therefore that the central value between adjacent cells is taken (central scheme). A (first-order) **upwind scheme** gives more weight to the upstream cell. This can increase stability of advective transport simulation, but may increase numerical diffusion (see Pe-criterion below). If these cell indices are *in* (*i*, or *i*–1) and *out* (*i* or *i*–1), respectively, depending on the water flow direction, the heat flux can be written as

$$J_{x,i,j,k,\text{adv}}(t^*) = q_{x,i-\frac{1}{2},j,k}(t^*)C_w T_{\text{in},j,k}(t^*)\Delta y \Delta z$$

$$- q_{x,i+\frac{1}{2},j,k}(t^*)C_w T_{\text{out},j,k}(t^*)\Delta y \Delta z \qquad (4.19)$$

The **conductive heat flux** into the cell is expressed as

$$J_{x,i,j,k,\text{diff}}(t^*) = \frac{\lambda_{x,i-\frac{1}{2},j,k} \cdot \left(T_{i,j,k}(t^*) - T_{i-1,j,k}(t^*)\right)}{\Delta x}\Delta y \Delta z$$

$$- \frac{\lambda_{x,i-\frac{1}{2},j,k} \cdot \left(T_{i+1,j,k}(t^*) - T_{i,j,k}(t^*)\right)}{\Delta x}\Delta y \Delta z \qquad (4.20)$$

In a corresponding manner, the **dispersive flux** in x-direction caused by a temperature gradient in x-direction is

$$J_{x,\text{disp},xx}(t^*) = \frac{D_{xx,i-\frac{1}{2},j,k} \cdot \left(T_{i,j,k}(t^*) - T_{i-1,j,k}(t^*)\right)}{\Delta x} \Delta y \Delta z$$

$$- \frac{D_{xx,i-\frac{1}{2},j,k} \cdot \left(T_{i+1,j,k}(t^*) - T_{i,j,k}(t^*)\right)}{\Delta x} \Delta y \Delta z \qquad (4.21)$$

The **dispersive flux in x-direction caused by a temperature gradient in y-direction,** using interpolation of the gradient to the center of the exchanging interface, is

$$J_{x,\text{disp},xy}(t^*) = \frac{C_w D_{xx,i-\frac{1}{2},j,k} \cdot \left(T_{i-1,j-1,k}(t^*) - T_{i-1,j+1,k}(t^*) + T_{i,j-1,k}(t^*) - T_{i,j+1,k}(t^*)\right)}{4\Delta y} \Delta y \Delta z$$

$$- \frac{D_{xx,i-\frac{1}{2},j,k} \cdot \left(T_{i+1,j,k}(t^*) - T_{i,j,k}(t^*)\right)}{\Delta x} \Delta y \Delta z \qquad (4.22)$$

In a similar manner, all dispersive fluxes into the cell can be expressed. In order to express the dispersive fluxes in a symmetric way, a total of 26 neighboring cells are needed in three dimensions.

Finally, the **heat storage** in the cell (i, j, k) is

$$\frac{C_m \cdot \left(T_{i,j,k}(t + \Delta t) - T_{i,j,k}(t)\right) \Delta x \Delta y \Delta z}{\Delta t}$$

$$= \sum_{i=x,y,z} \left(J_{i,\text{adv}}(t^*) + J_{i,\text{diff}}(t^*) + \sum_{j=x,j,z} J_{i,\text{disp}\_ij}(t^*) \right) + P_{t,i,j,k}(t^*) \qquad (4.23)$$

Still, the time t^*, at which the fluxes are determined, remains to be specified.

If all fluxes are evaluated at the old time level, that is, $t^* = t$, one obtains an **explicit scheme.** This formulation is attractive since all fluxes can be evaluated directly without the need to solve an equation system. The only unknown is the temperature $T_{ijk}(t + \Delta t)$ appearing in the storage term, which can be computed in an explicit manner. However, the time step needs to be small enough in order to guarantee stability. It is required that the diffusive and dispersive thermal fluxes into any cell are smaller than the rate of change of energy within the cell:

$$2 \cdot (\lambda_{\mathrm{m}} + C_w D_{\mathrm{t,disp,L}}) \frac{T}{\Delta s} \leq \frac{C_{\mathrm{m}} T \Delta s}{\Delta t} \tag{4.24}$$

This leads to the **von Neumann criterion for the time step** Δt of explicit schemes, which states

$$\Delta t \leq \frac{C_{\mathrm{m}} \Delta s^2}{2 \cdot (\lambda_{\mathrm{m}} + C_w D_{\mathrm{t,disp,L}})} = \frac{\Delta s^2}{2 D_t} \tag{4.25}$$

where Δs is the **spatial discretization** (Δx, or Δy, or Δz).

Using $t^* = t + \Delta t$ yields a fully **implicit scheme**, while for a **Crank–Nicolson scheme**, $T(t^*) = 0.5 \cdot (T(t + \Delta t) + T(t))$ (time-centered) is applied (which also leads to implicit equations). Very often in available codes, the difference scheme is an input parameter and has to be selected: upwind or central in space, and explicit, implicit, or centered in time. For the implicit and the time-centered schemes, a **linear equation system** is obtained, of the form

$$\sum_{j=1}^{n} A_{ij} T_j = b_i; \quad i = 1, \ldots, n \tag{4.26}$$

where A_{ij} is a term of the coefficient matrix, T_j is a component of an unknown vector of the temperature in the cells, b_i is a constant term, and n is the number of cells. The resulting **matrix** [A] is sparse, since only the neighboring cells are involved in the balance equations. In contrast to the flow equations, in general, matrix [A] is **not symmetric**. The reason lies in the advective terms. Adapted numerical techniques, like the biconjugate gradient solver, have to be used.

Still accuracy **criteria** with respect to the choice of Δs and Δt have to be observed by the modeler in order to avoid excessive numerical diffusion (also called numerical dispersion) and numerical oscillations.

In order to **reduce numerical diffusion**, which means that the discretization is able to adequately represent sharp thermal fronts without smearing, it is required that the advective heat flux is smaller than the dispersive and diffusive heat fluxes everywhere within the solution domain. For extreme cases, this requirement can be stated heuristically as

$$C_w q \frac{T}{2} \leq (\lambda_{\mathrm{m}} + C_w D_{\mathrm{t,disp,L}}) \frac{T}{\Delta s} \tag{4.27}$$

This leads to the **requirement for the spatial discretization** Δs:

$$\Delta s \leq \frac{2 \cdot \left(\dfrac{\lambda_m}{C_w} + D_{t,\mathrm{disp,L}} \right)}{q} = \frac{2 D_t}{q} \tag{4.28}$$

For **dominant macrodispersion**, the criterion reduces to

$$\Delta s \leq 2\beta_L \tag{4.29}$$

An equivalent formulation is that the **thermal grid Peclet number** satisfies the following:

$$\mathrm{Pe} = \frac{q\Delta s}{D_t} \leq 2 \tag{4.30}$$

In order to **reduce numerical oscillations** of the solution, which manifest themselves as overshooting and undershooting, it is required that the advective heat flux into any cell is smaller than the rate of change of energy within the cell. For extreme cases, this requirement can be stated heuristically as

$$C_w q T \leq \frac{C_m T \Delta s}{\Delta t} \tag{4.31}$$

This leads to the **requirement for the time step**:

$$\Delta t \leq \frac{C_m \Delta s}{C_w q} \tag{4.32}$$

An equivalent formulation is that the **thermal grid Courant number** Co satisfies the **Courant–Friedrichs–Lewy stability condition**:

$$\mathrm{Co} = \frac{C_w q \Delta t}{C_m \Delta s} \leq 1 \tag{4.33}$$

Compared to solute transport, the criterion is relaxed by the thermal retardation factor.

Still **lateral numerical diffusion** has to be controlled. From experience with solute plume simulations, it is desirable to resolve the thermal plume laterally by at least 10 cells.

Finally, it is recommended to check whether the numerical results are **grid convergent**. This means that the results using a finer grid are practically invariant compared to the original grid. An **application** of the finite difference method is given, for example, in Birkholzer and Tsang (2000).

4.2.2 Principles of the finite element method for heat transport

In the finite element method, discretization uses the so-called **finite elements** (e.g., prismatic elements; Figure 4.7). Various classes of finite elements are available. Each element contains a number of **nodal points,** where the approximate solution is sought. The simplest two-dimensional and three-dimensional finite elements are triangular (3 nodal points) and tetrahedral (4 nodal points) elements with linear interpolation functions. Due to their flexible shape, finite element grids can much better adapt to irregular boundaries of the domain D and can also be better refined locally in regions where better resolution is needed, for example, close to sources and sinks. Within the solution domain, a **trial solution** is defined as follows:

$$\hat{T}(\mathbf{x},t) = \sum_{i=1}^{n} T_i(t)w_i(\mathbf{x}) \tag{4.34}$$

The functions $w_i(\mathbf{x})$ are **weighting functions**, which are determined from **the interpolation functions** of the finite element, and n is the number of nodal points. In fact, $w_i(\mathbf{x})$ vanishes outside the neighborhood of a nodal point (element patch containing all elements connected to the nodal point).

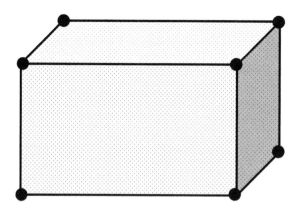

Figure 4.7 Prismatic finite element with nodal points.

The parameters $T_i(t)$ are the **unknown nodal values**. The trial solution approximates the temperature distribution within the domain. Inserting the trial solution into the rearranged heat transport equation:

$$L(\hat{T}) = \nabla \cdot \left[\mathbf{D}_t \nabla \hat{T} \right] - \frac{C_w}{C_m} \nabla \cdot (\mathbf{q}\hat{T}) + \frac{P_t}{C_m} - \frac{\partial \hat{T}}{\partial t} = \varepsilon(\mathbf{x}, t) \tag{4.35}$$

yields a **residual** $\varepsilon(\mathbf{x}, t)$. According to **Galerkin**, it is required that the **weighted residual** using $w_i(\mathbf{x})$ as a weighting function **vanishes in the neighborhood of all nodal points** i with unknown variable T:

$$\int_D \varepsilon(\mathbf{x}, t) w_i(\mathbf{x}) \, dD = 0; \quad i = 1, \dots n \tag{4.36}$$

Integration is, in fact, restricted to the element patch of each nodal point. Therefore, the condition states that the weighted residual vanishes in the neighborhood of each nodal point. To avoid the appearance of distributions, one integration step of the conduction-thermal dispersion term is carried over to the weighting function using Greene's theorem. Integration (analytical if possible or numerical using Gauss points) yields a **set of ordinary differential equations** as follows:

$$\sum_{j=1}^{n} F_{ij}(t^*) \left(\frac{dT_j}{dt} \right) + \sum_{j=1}^{n} A_{ij}(t^*) T_j = b_i; \quad i = 1, \dots, n \tag{4.37}$$

where A_{ij} and F_{ij} are matrix elements and b_i is a constant term. Still it has to be decided for which time t^* the matrices are evaluated. For a fully **implicit scheme** $t^* = t + \Delta t$, and for a **Crank–Nicolson scheme** $T(t^*) = 0.5 \cdot (T(t + \Delta t) + T(t))$. Since the matrix [F] is not a diagonal matrix, an explicit scheme does not provide a computational advantage, unless [F] is diagonalized by lumping all terms of a row in the diagonal. The **time derivative** is often discretized in finite differences with:

$$\frac{dT_j}{dt} \simeq \frac{T_j(t + \Delta t) - T_j(t)}{\Delta t} \tag{4.38}$$

leading to a linear equation system for the unknown nodal values $T_j(t + \Delta t)$ of the temperature. The resulting **matrix [A]** is again sparse, since only the neighboring nodal points are involved in the integration. Again, in general, the matrix [A] is **not symmetric** due to the advective terms. Adapted numerical techniques, like the biconjugate gradient solver, have to be used.

An alternative consists of using an explicit scheme for the advective term and an implicit scheme for the diffusive–dispersive term, which leads to a symmetric matrix [A] (e.g., Leismann and Frind 1989).

Finite element techniques exhibit similar **numerical stability problems** as the finite difference methods. Again, the **grid Peclet number** (Equation 4.30) and the **Courant number** (Equation 4.33) **criteria** have to be observed. In order to avoid lateral numerical diffusion, finite element grids can be aligned along water flow lines leading to the principal direction technique (Frind and Germain 1986). This avoids lateral numerical diffusion. An **application** of the finite element method is given, for example, in Molson et al. (1992).

4.2.3 Principles of the finite volume method for heat transport

In the finite volume method, the solution domain is divided into small (convex) **finite volumes**. Nodal points are used to interpolate the field variable. Usually a single node is used for each finite volume. **Surface heat fluxes** are expressed either by Gauss' divergence theorem or directly by approximating the fluxes. The sum of all inflowing heat fluxes is equal to the rate of change of energy within the finite volume according to heat conservation. Boundary fluxes (Neumann type boundary conditions) can be directly introduced into the balance equation. **For grids using rectangular blocks as finite volumes, the approach is identical to the finite difference approach** as described in Section 4.2.1. However, the finite volume method is **more general** and allows unstructured grids. A model formulation using the finite volume method can be found in Russell et al. (2003). **Applications** of the finite volume method are given in Clauser (2003) and Rühaak et al. (2008).

4.2.4 Principles of the method of characteristics for heat transport

In the **method of characteristics** used for transport problems in groundwater, the transport step is split into two half steps (operator splitting), one purely advective and the other diffusive and dispersive. In a **moving control volume**, moving with the thermal front velocity $\mathbf{u}_t$, the change of temperature can be expressed as follows:

$$
\begin{aligned}
\frac{dT}{dt} &= \frac{\partial T}{\partial t} + \frac{\partial T}{\partial x}\frac{\partial x}{\partial t} + \frac{\partial T}{\partial y}\frac{\partial y}{\partial t} + \frac{\partial T}{\partial z}\frac{\partial z}{\partial t} \\
&= \frac{\partial T}{\partial t} + \frac{\partial T}{\partial x}u_{t,x} + \frac{\partial T}{\partial y}u_{t,y} + \frac{\partial T}{\partial z}u_{t,z}
\end{aligned}
\tag{4.39}
$$

The resulting tracks $x(t)$, $y(t)$, and $z(t)$ of the particles are called the **characteristics**. The **diffusive and dispersive half time step** is treated with the conventional finite difference method, whereas the **advection half time step** is performed using **particle tracking** based on the velocity field. A large number of particles are initially introduced in the solution domain. Each particle carries a temperature, which can change over time. The **principal steps** are as follows:

- Given temperature $T_p(t)$ of particles and $T_{ij}(t)$ of cells.
- All particles are moved advectively. New intermediate cell temperatures $T_{ij}^*(t + \Delta t)$ are calculated by averaging the particle temperatures in the cells.
- Purely diffusive and dispersive temperature changes are calculated on the finite difference grid using $T_{ij}^*(t + \Delta t)$, yielding new cell temperatures $T_{ij}(t + \Delta t)$.
- New particle temperature values $T_p(t + \Delta t)$ are calculated by adding increments.
- Start a new time step.

It is important that an optimal **interpolation of the velocity** within the cells is obtained. Frequently used interpolation schemes follow Prickett et al. (1981) and Pollock (1988). Modified (e.g., Liu and Dane 1996) and hybrid (forward and backward) schemes of the method of characteristics do exist. Numerical oscillations of the solution occur due to the particle-based nature of the method. Increasing the number of particles in the system can reduce these oscillations.

An **application** is presented, for example, in Hecht-Méndez et al. (2010).

4.2.5 Principles of the random walk method for heat transport

In the **random walk method** used for transport problems in groundwater, the transport step is again split into two half steps, one purely advective and the other diffusive and dispersive as in the method of characteristics. Again, the **advective time step** is performed by **particle tracking**, based on the velocity field. In contrast to the method of characteristics, the particles carry a fixed energy in the random walk method. A large ensemble of particle paths yields, in the limit, the solution of the transport equation. From the analytical solution of the transport problem for uniform flow velocity in the direction of x, the **new particle position** after a time step Δt, starting at the location $x(t = 0) = x_0$, is

$$x_p\left(t + \Delta t\right) = x_p\left(t\right) + u_{t,x}\,\Delta t + Z\sqrt{2\,D_{t,L}\,\Delta t} \qquad (4.40)$$

where Z is a normally distributed random number with zero average and standard deviation $\sigma_Z = 1$. The y- and z-components are obtained in a similar manner using transversal coefficients. The method yields a distribution of particles of equal energy. The cell temperature is obtained by counting the particles within a cell and dividing the total energy by the total heat capacity of the aquifer volume in the cell. Therefore, enough particles have to be added to the system. Since Equation 4.40 is based on a uniform velocity field, which is not the general case, **correction terms** have to be added (referred to as Fokker–Planck terms). The **new particle position** is for two-dimensional heat transport:

$$x_p(t + \Delta t) = x_p(t) + \left[u_{t,x} + \frac{\partial D_{t,xx}}{\partial x} + \frac{\partial D_{t,xy}}{\partial y} \right]_{(x_p(t))} \cdot \Delta t + Z' \cdot \left[\sqrt{2D_{t,L}\Delta t}\, \frac{u_{t,x}}{u_t} \right]_{(x_p(t))}$$

$$- Z'' \cdot \left[\sqrt{2D_{t,T}}\, t\, \frac{u_{t,y}}{u_t} \right]_{(x_p(t))}$$

$$y_p(t + t) = y_p(t) + \left[u_{t,y} + \frac{\partial D_{t,yx}}{\partial x} + \frac{\partial D_{t,yy}}{\partial y} \right]_{(x_p(t))} \cdot \Delta t + Z' \cdot \left[\sqrt{2D_{t,L}\Delta t}\, \frac{u_{t,y}}{u_t} \right]_{(x_p(t))}$$

$$+ Z'' \cdot \left[\sqrt{2D_{t,T}\Delta t}\, \frac{u_{t,x}}{u_t} \right]_{(x_p(t))} \tag{4.41}$$

where u_t is the absolute value of the thermal velocity, and Z' as well as Z'' are normally distributed random numbers (zero mean and unit standard deviation). The correction term prevents, for example, particles from accumulating at stagnation points (Kinzelbach 1987). It is again important that an optimal **interpolation of the velocity** within the cells is obtained. Frequently used interpolation schemes are after Prickett et al. (1981) and after Pollock (1988). For continuous injection of heat, particles have to be added continuously. A three-dimensional formulation for solute transport by random walk can be found in Kinzelbach and Uffink (1991) and Lichtner et al. (2002). A two-dimensional **application** in heat transport is presented in Chevalier and Banton (1999).

4.3 STRATEGY FOR COUPLED FLOW AND HEAT TRANSPORT

Usually, **coupled flow and heat transport equations** are linearized using a **Picard iteration scheme** (point iteration). In a first step, the flow equation is solved. Based on the head values, the Darcy velocity field is evaluated in a second step. In a third step, the heat transport equation is solved, and in a fourth step, the temperature-dependent parameters like water density are updated. All four steps are iterated to a specified convergence tolerance.

Various codes are organized in this manner (see Ackerer et al. 2004; Molson and Frind 2012). Molson et al. (1992) handle nonlinearities in the coupled flow and heat transport equations by centering the nonlinear terms in time during the iteration. In order to speed up the iteration process, Ackerer et al. (2004) suggest that the heat transport equation is evaluated first, and then an update of the temperature-dependent parameters is performed followed by solving the flow equation with the evaluation of the velocity field. In their scheme, the heat transport equation is solved with velocities defined in the previous iteration. Because the flow equation is more dependent on the temperature than the heat transport equation on the heads (temperature variations in time create a sink/source term in the flow equation), this algorithm should reduce the number of iterations needed within one time step. However, the **solution of highly nonlinear, density-dependent flow problems** involving high temperature contrasts may require other solution approaches (see, e.g., Herbert et al. 1988).

Diersch and Kolditz (1998) use **first- or second-order predictor–corrector schemes** for solving the coupled equations.

4.4 SOME AVAILABLE CODES FOR THERMAL TRANSPORT MODELING IN GROUNDWATER

As already stated, a **large number of codes on solute and contaminant transport** are available, such as MT3D (Zheng 1990) or MT3DMS (Zheng and Wang 1999), which, in principle, can also be used for thermal transport studies in groundwater by analogy. Among the codes that directly allow simulation of thermal processes, we mention a few. A list of some selected groundwater flow and heat transport codes are presented in Anderson (2005).

FEFLOW (DHI-WASY 2010) is a variably saturated, three-dimensional finite element code for the simulation of variable-density water flow, solute, and heat transport, including coupled transport. Thermal applications are presented, for example, in Maréchal et al. (1999) and Nam et al. (2008). The software includes analytical and numerical modules for the finite element formulation of BHEs in modeling geothermal heating systems.

HST3D (Kipp 1997) is a three-dimensional finite difference code. It simulates groundwater flow and associated heat and solute transport in saturated aquifers. It can handle variable water density and viscosity. Among other applications, the code is offered for heat storage in aquifers. An application using the code is given in Bravo et al. (2002).

HEATFLOW-SMOKER (Molson and Frind 2012) is a three-dimensional finite element code for solving complex density-dependent groundwater flow and thermal energy transport problems. The model can be used to solve one-, two-, or three-dimensional heat transport problems within a variety of hydrogeological systems, including discretely fractured porous media.

HydroGeoSphere (Graf and Therrien 2007; Raymond et al. 2011) is a three-dimensional numerical model for fully integrated density-dependent subsurface and surface flow, heat transport, and solute transport. It was used by Raymond et al. (2011) for a numerical analysis of thermal response tests. A review was presented by Brunner and Simmons (2012).

SEAWAT (Langevin et al. 2007) is a coupled version of MODFLOW and MT3DMS models designed to simulate three-dimensional, variable-density, groundwater flow and solute transport in saturated porous media. The effects of fluid viscosity variation on groundwater flow are included. Although not explicitly designed to model heat transport, temperature can be simulated as one of the species by entering appropriate transport coefficients. Version 4 is based on MODFLOW-2000 and MT3DMS. Applications are presented, for example, in Vandenbohede and Lebbe (2011).

SHEMAT (Clauser 2003) is a three-dimensional finite difference code. It mainly focuses on numerical simulation of reactive flow in geothermal aquifers. It solves transient coupled problems of groundwater flow, heat transport, species transport, and chemical water–rock interaction in fluid-saturated porous media. Applications are presented, for example, in Clauser (2003) and Pannike et al. (2006).

SPRING (delta-h 2012) is a variably saturated three-dimensional finite element code for the simulation of coupled water flow and solute or heat transport in saturated and unsaturated porous media. An application is shown, for example, in Engeler et al. (2011).

SUTRA (Voss and Provost 2010) is a variably saturated three-dimensional finite element code for density-dependent saturated or unsaturated groundwater flow, and solute or heat transport. An application is presented, for example, in Ronan et al. (1998).

TOUGH2 (Pruess et al. 2012) is a variably saturated three-dimensional integral finite difference code for nonisothermal flows of multicomponent, multiphase fluids and coupled heat transfer in one-, two-, and three-dimensional porous and fractured media. Temperature and pressure dependence of thermophysical properties are taken into account. Main applications for which TOUGH2 was designed are in geothermal reservoir engineering, nuclear waste disposal, environmental assessment and remediation, and unsaturated and saturated zone hydrology. A thermal application is presented, for example, in Birkholzer and Zhang (2000).

VS2DI (Hsieh et al. 2000) is a two-dimensional finite difference code for simulating fluid flow and solute or heat transport in variably saturated porous media in one or two dimensions using Cartesian or radial coordinate systems. An application is given, for example, in Constantz (1998).

Table 4.1 gives an **overview on various codes** that are suited for heat transport simulations of shallow geothermal systems considering groundwater flow.

Table 4.1 Numerical codes suitable for heat transport simulations of shallow geothermal systems considering groundwater flow (not meant to be exhaustive or complete)

Code name	Numerical method	Processes	Coupling	Availability	Comments	Reference
AST/TWOW	FD	H,T	H → T	Commercial	3D, calculates near-field heat transport around BHEs	Schmidt and Hellström (2005)
BASIN2	FD	H,T	H ↔ T, M, CH	Free code	2D, simulates sedimentary basin development; cross-sectional view	Bethke et al. (2007)
COMSOL	FE	H,T,C	H ↔ T	Commercial	3D, multiphysics (more processes can be coupled)	Holzbecher and Kohfahl (2008)
FEFLOW	FE	H,T,C	H ↔ T, M, C	Commercial	2D, 3D	DHI-WASY (2010)
FRACHEM	FE	H,T,C	H ↔ T, M, C	Scientific	3D, used for hot dry rock modeling	Bächler (2003)
FRACture	FE	H,T	H ↔ T, M	Scientific	3D, developed for hot dry rock modeling	Kohl and Hopkirk (1995)
ROCKFLOW/GeoSys	FE	H,T,C	H ↔ T, C	Scientific	3D, fracture systems can be included. Allows for multiphase flow	Kolditz et al. (2001)
HEATFLOW-SMOKER	FE	H,T	H ↔ T	Free code	1D, 2D, 3D	Molson and Frind (2012)
HST2D/3D	FD	H,T,C	H ↔ T, M, R	Free code	2D, 3D	Kipp (1997)

HydroTherm	FE	H,T	H↔T	Free code	2D, 3D, two-phase model; can simulate 0 to 1200°C	Kipp et al. (2008)
HydroGeoSphere	FV	H,T,C	H↔T,C	Scientific	3D	Raymond et al. (2011)
HYDRUS-2D/3D	FE	H,T,C	H→T	Commercial	Unsaturated zone, plant water uptake is considered	Radcliffe and Šimůnek (2010)
SEAWAT	FD	H,T,C	H↔T,C	Free code	3D	Langevin et al. (2008)
SHEMAT	FD	H,T,C	H↔T,C	Commercial	3D	Clauser (2003)
SUTRA	FE	H,T,C	H↔T,C	Free code	2D, 3D	Voss and Provost (2010)
SPRING	FE	H,T,C	H↔T,C	Commercial	3D	delta-h (2012)
THETA	FD	H,T,C	H↔T,C	Scientific	3D	Kangas (1996)
TOUGH2	FD	H,T,C	H↔T, M, R	Commercial	1D, 2D, and 3D, one of the most widely used codes in geothermal energy technologies; allows for multiphase flow	Pruess et al. (2012)
TRADIKON 3D	FD	H,T	H→T	Free code	3D, specially designed for BHE assessments	Brehm (1989)
VS2DI	FD	H,T	H→T	Free code	2D	Hsieh et al. (2000)

Source: After Hecht-Méndez, J. et al. *Ground Water* 48(5), 741–756, 2010.

Notes: C, contaminant (solute); H, hydraulic; H → T, fluid flow is independent of T; H ↔ T, fluid flow depends on T; T, temperature; M, mechanical deformation (pore deformation); R, chemical reaction.

REFERENCES

Ackerer, P., Younès, A., Mancip, M. (2004). A new coupling algorithm for density driven flow in porous media. *Geophysical Research Letters* 31, L12506, doi:10.1029/2004GL019496.

Al-Khoury, R. (2012). *Computational Modeling of Shallow Geothermal Systems.* CRC-Press/Baalkema, Taylor and Francis Group, Boca Raton, USA.

Al-Khoury, R., Bonnier, P.G., Brinkgreve, R.B.J. (2005). Efficient finite element formulation for geothermal heating systems. Part I: Steady-state. *International Journal for Numerical Methods in Engineering* 63, 988–1013.

Al-Khoury, R., Bonnier, P.G. (2006). Efficient finite element formulation for geothermal heating systems. Part II: Transient. *International Journal for Numerical Methods in Engineering* 67, 725–745.

Al-Khoury, R., Köbel, T., Schramedei, R. (2010). Efficient numerical modeling of heat exchangers. *Computers and Geosciences* 36, 1301–1315.

Anderson, M.P. (2005). Heat as a groundwater tracer. *Ground Water* 43(6), 951–968.

Bächler, D. (2003). Coupled thermal-hydraulic-chemical modeling at the Soultz-sous-Forêts HDR Reservoir (France). Ph.D. diss., Swiss Federal Institute of Technology, Zurich.

Bauer, D., Heidemann, W., Müller-Steinhagen, H., Diersch, H.-J.G. (2011). Thermal resistance and capacity models for borehole heat exchangers. *International Journal of Energy Research* 35, 312–320.

Bethke, C., Lee, M.-K., Park, J. (2007). *Basin Modeling with Basin2, Release 5.0.1.* Hydrogeology Program, University of Illinois, Urbana Champaign, Illinois.

Birkholzer, J.T., Zhang, Y.W. (2000). Modeling the thermal-hydrologic processes in a large-scale underground heater test in partially saturated fractured tuff. *Water Resources Research* 36(6), 1431–1447.

Bravo, H.R., Feng, J., Hunt, R.J. (2002). Using groundwater temperature data to constrain parameter estimation in a groundwater flow model of a wetland system. *Water Resources Research* 38(8), 28-1–28-14, doi:10.1029/2000WR000172.

Brehm, D. (1989). *Development, Validation and Application of a 3-Dimensional, Coupled Flow and Transport Finite Differences Model.* Giessener Geologische Schriften, Giessen, Lenz Verlag.

Brookfield, A.E., Sudicky, E.A., Parks, Y.-J., Conant, B. (2009). Thermal transport modelling in a fully integrated surface/subsurface framework. *Hydrological Process* 23(15), 2150–2164.

Brunner, P., Simmons, C.T. (2012). HydroGeoSphere: A fully integrated, physically based hydrological model. *Ground Water* 50(2), 170–176.

Chevalier, S., Banton, O. (1999). Modelling of heat transfer with the random walk method. Part 1. Application to thermal energy storage in porous aquifers. *Journal of Hydrology* 222, 129–139.

Chiang, W.-X., Kinzelbach, W. (2005). *2D Groundwater Modeling with PMWIN: A Simulation System for Modeling Flow and Transport Processes.* Springer, Berlin, 2005.

Chiasson, A.D., Rees, S.J., Spitler, J.D. (2000). A preliminary assessment of the effects of groundwater flow on closed-loop ground source heat pump systems. *ASHRAE Transactions* 106(1), 380–393.

Clauser, C. (Ed.) (2003). *Numerical Simulation of Reactive Flow in Hot Aquifers—SHEMAT and Processing SHEMAT*. Springer, Berlin.

Constantz, J. (1998). Interaction between stream temperature, streamflow, and groundwater exchanges in alpine streams. *Water Resources Research* 34(7), 1609–1616.

Delta-h (2012). *SPRING 4 Online Manual*. Delta-h, Witten, Germany.

Deng, Z., Rees, S.J., Spitler, J.D. (2005). A model for annual simulation of standing column well ground heat exchangers. *HVAC&R Research* 11, 637–656.

DHI-WASY (2010). *FEFLOW 6. User Manual*. DHI-WASY GmbH, Berlin, Germany.

Diersch, H.-J.G. (1996). Interactive, graphics-based finite-element simulation system FEFLOW for modeling groundwater flow, contaminant mass and heat transport processes. User's Manual Version 4.5, April 1996, WASY Institute for Water Resources Planning and Systems Research Ltd, Berlin. DHI-WASY GmbH, DHI-WASY GmbH, Berlin, Germany.

Diersch, H.-J., Kolditz, O. (1998). Coupled groundwater flow and transport: 2. Thermohaline and 3D convection systems. *Advances in Water Resources* 21, 401–425.

Diersch, H.-J.G., Bauer, D., Heidemann, W., Rühaak, W., Schätzl, P. (2011a). Finite element modeling of borehole heat exchanger systems. Part I Fundamentals. *Computers and Geosciences* 37, 1122–1135.

Diersch, H.-J.G., Bauer, D., Heidemann, W., Rühaak, W., Schätzl, P. (2011b). Finite element modeling of borehole heat exchanger systems. Part II. Numerical simulations. *Computers and Geosciences* 37, 1136–1147.

Doughty, C., Hellström, G., Tsang, C.F., Claesson, J. (1982). A dimensionless parameter approach to the thermal behaviour of an aquifer thermal energy storage system. *Water Resources Research* 18(3), 571–587.

Dwyer, T.E., Eckstein, Y. (1987). Finite-element simulation of low-temperature, heat-pump-coupled, aquifer thermal energy storage. *Journal of Hydrology* 95, 19–38.

Engeler, I., Hendricks Franssen, H.J., Müller, R., Stauffer, F. (2011). The importance of coupled modelling of variably saturated groundwater flow-heat transport for assessing river-aquifer interactions. *Journal of Hydrology* 397, 295–305.

Fan, R., Jiang, J., Yao., Y., Shiming, D., Ma, Z. (2007). A study of the performance of a geothermal heat exchanger under coupled heat conduction and groundwater advection. *Energy* 32, 2199–2209.

Ferguson, G. (2007). Heterogeneity and thermal modeling in ground water. *Ground Water* 45(4), 485–490.

Frind, E.O., Germain, D. (1986). Simulation of contaminant plumes with large dispersive contrast: Evaluation of alternating direction Galerkin models. *Water Resources Research* 22(13), 1857–1873.

Fujimitsu, Y., Fukuoka, K., Ehara, S., Takeshita, H., Abe, F. (2010). Evaluation of subsurface thermal environment change caused by a ground-coupled heat pump system. *Current Applied Physics* 11 (1) Supplement S113–S116.

Gao, Q., Zhou, X.-Z., Jiang, J., Chen, X.-L., Yan, Y.-Y. (2013). Numerical simulation of the thermal interaction between pumping and injecting well groups. *Applied Thermal Engineering* 51, 10–19.

Glück, B. (2011). Simulationsmodell Erdwärmesonden zur wärmetechnischen Beurteilung von Wärmequellen, Wärmesenken und Wärme/Kältespeichern (in German) (Simulation model for BHE for the assessment of heat sources and sinks and heat storage). www.berndglueck.de.

Graf, T., Therrien, R. (2007). Coupled thermohaline groundwater flow and single-species reactive solute transport in fractured porous media. *Advances in Water Resources* 30, 742–771.

Hecht-Méndez, J., Molina-Giraldo, N., Blum, P., Bayer, P. (2010). Evaluating MT3DMS for heat transport simulation of closed geothermal systems. *Ground Water* 48(5), 741–756.

Herbert, A.W., Jackson, C.P., Lever, D.A. (1988). Coupled groundwater flow and solute transport with fluid density strongly dependent upon concentration. *Water Resources Research* 24(10), 1781–1795.

Hidalgo, J.J., Carrera, J., Dentz, M. (2009). Steady state heat transport in 3D heterogeneous porous media. *Advances in Water Resources* 32(8), 1206–1212.

Holzbecher, E., Kohfahl, C. (2008). *Geothermics Modelling using COMSOL Multiphysics*. Seminar manual. Berlin.

Hsieh, P.A., Wingle, W., Healy, R.W. (2000). VS2DI-A graphical software package for simulating fluid flow and solute or energy transport in variably saturated porous media. USGS Water Resources Investigations Report 9, 9-4130. USGS, Denver, Colorado.

Jalaluddin, Miyara, A. (2012). Thermal performance investigation of several types of vertical ground heat exchangers with different operation mode. Applied Thermal Engineering 33–34, p. 167–174. doi: 10.1016/j.applthermaleng.2011.09.030.

Kangas, M.T. (1996). Modeling of transport processes in porous media for energy applications. Ph.D. thesis, Helsinki University of Technology, Helsinki, Finland.

Kim, J., Lee, Y., Yoon, W.S., Yeon, Y.S., Koo, M.-H., Keehm, Y. (2010). Numerical modeling of aquifer thermal energy storage systems. *Energy* 35, 4955–4965.

Kinzelbach, W. (1987). *Numerische Methoden zur Modellierung des Transports von Schadstoffen im Grundwasser. Schriftenreihe GWF Wasser-Abwasser*. Vol. 21, Oldenburg Verlag, Munich, Germany.

Kinzelbach, W., Uffink, G. (1991). The random walk method and extensions in groundwater modelling. In: Transport Processes in Porous Media, J. Bear and M.Y. Corapcioglu (eds.), Kluwer Academic Publishers, Dordrecht, The Netherlands, pp. 761–787.

Kipp, K.L. (1997). Guide to the revised heat and solute transport simulator: HST3D—Version 2. Water-Resources Investigations Report 97-4157, USGS, Denver, Colorado.

Kipp, K.L., Jr., Hsieh, P.A., Charlton, S.R. (2008). Guide to the revised groundwater flow and heat transport simulator: HYDROTHERM—Version 3: U.S. Geological Survey Techniques and Methods 6-A25, 160 p.

Kohl, T., Hopkirk, R.J. (1995). FRACTure—A simulation code for forced fluid flow and transport in fractured rock. *Geothermics* 24(3), 333–343.

Laloui, L., Nuth, M., Vulliet, L. (2006). Experimental and numerical investigations of the behavior of a heat exchanger pile. *International Journal for Numerical and Analytical Method in Geomechanics* 30, 763–781.

Langevin, C.D., Thorne, D.T., Jr., Dausman, A.M., Sukop, M.C. Guo, W. (2008). SEAWAT Version 4: A Computer program for simulation of multi-species solute and heat transport. *U.S. Geological Survey Techniques and Methods*. Book 6, Chap. A22. USGS, Reston, Virginia.

Lazzari, S., Priarone, A., Zanchini, E. (2010). Long-term performance of BHE (borehole heat exchanger) with negligible groundwater flow. *Energy* 35, 4966–4974.

Lee, C.K., Lam, H.N. (2008). Computer simulation of borehole ground heat exchangers for geothermal heat pump systems. *Renewable Energy* 33, 1286–1296.

Leismann, H.M., Frind, E.O. (1989). A symmetric-matrix time integration scheme for the efficient solution of advection-dispersion problems. *Water Resources Research* 25(6), 1133–1139.

Lichtner, P.C., Kelkar, S., Robinson, B. (2002). New form of dispersion tensor for axisymmetric porous media with implementation in particle tracking. *Water Resources Research* 38(8), 1146, doi:10.1029/2000WR000100.

Lippmann, M.J., Tsang, C.F., Witherspoon, P.A. (1977). Analysis of the response of geothermal reservoirs under injection and production procedures. SPE Paper No. 6537.

Liu, H.H., Dane, J.H. (1996). An interpolation-corrected modified method of characteristics to solve advection-dispersion equations. *Advances in Water Resources* 19(6), 359–368.

Maréchal, J.C., Perrochet, P., Tacher, L. (1999). Longterm simulations of thermal and hydraulic characteristics in a mountain massif: The Mont Blanc case study, French and Italian Alps. *Hydrogeology Journal* 7(4), 341–354.

Mercer, J.W., Faust, C.R., Miller, W.J., Pearson, F.J., Jr. (1982). Review of simulation techniques for aquifers thermal energy storage (ATES). *Advances in Hydroscience* 13, 1–129. Academic Press, New York.

Mercer, J.W., Pinder, G.F., Donaldson, I.G. (1975). A Galerkin finite element analysis of the hydrothermal system at Wairakei, New Zealand. *Journal of Geophysical Research* 80(17), 2608–2621.

Merheb, F. (1984). Modèle de gestion des echanges hydrothermiques dans les nappes souterraines: Application à la région de Strasbourg. PhD Thesis Université Lous Pasteur de Strasbourg, Institut de Mécanique des Fluides.

Molson, J.W., Frind, E.O. (2012). HEATFLOW-SMOKER, Density-dependent flow and advective -dispersive transport of thermal energy, mass or residence time in three-dimensional porous or fractured porous media. User guide. Version 5.0, University of Laval and University of Waterloo, Canada.

Molson, J.W., Frind, E.O., Palmer, C.D. (1992). Thermal energy storage in an unconfined aquifer 2. Model development, validation, and application. *Water Resources Research* 28(19), 2857–2867.

Nam, Y., Ooka, R., Whang, S. (2008). Development of a numerical model to predict heat exchange rates for a ground-source heat pump system. *Energy and Buildings* 40, 2133–2140.

Painter, S., Seth, M.S. (2003). MULTIFLO User's Manual; MULTIFLO Version 2.0. Southwest Research Institute, San Antonio, Texas.

Pannike, S., Kölling, M., Panteleit, P., Reichling, J., Scheps, V., Schulz, H.D. (2006). Auswirkung hydrogeologischer Kenngrössen auf die Kältefahnen von Erdwärmesondenanlagen in Lockersedimenten. *Grundwasser* 11(1), 6–18.

Park, H., Lee, S.-R., Yoon, S., Choi, J.-C. (2013). Evaluation of thermal response and performance of PHC energy pile: Field experiments and numerical simulations. *Applied Energy* 103, 12–24.

Park, H., Lee, S.-R., Yoon, S., Shin, H., Lee, D.-S. (2012). Case study of heat transfer behavior of helical ground heat exchanger. *Energy and Buildings* 53, 137–144.

Parlange, M.B., Cahill, A.T., Nielsen, D.R., Hopmans, J.W., Wendroth, O. (1998). Review of heat and water movement in field soils. *Soil & Tillage Research* 47, 5–10.

Philip, J.R., de Vries, D.A. (1957). Moisture movement in porous materials under temperature gradients. *Transactions American Geophysical Union* 38, 222–232.

Pollock, D.W. (1988). Semianalytical computation of path lines of finite difference models. *Ground Water* 26(6), 743–750.

Prickett, T.A., Naymik, T.G., Lonnquist, C.G. (1981). A "random walk" transport model for selected groundwater quality evaluations. *Illinois State Water Survey Bulletin* 65.

Pruess, K., Oldenburg, C., Moridis, G. (2012). TOUGH2 User's Guide, Version 2.1. Report LBNL-43134 (revised), Lawrence Berkeley National Laboratory, Berkeley, California.

Radcliffe, D., Simůnek, J. (2010). *Soil Physics with HYDRUS. Modeling and Applications.* CRC Press, Boca Raton, Florida.

Raymond, J., Therrien, R., Gosselin, L., Lefebvre, R. (2011). Numerical analysis of thermal response tests with a groundwater flow and heat transfer model. *Renewable Energy* 36, 315–324.

Ronan, A.D., Prudic, D.E., Thodal, C.E., Constantz, J. (1998). Field study and simulation of diurnal temperature effects on infiltration and variably saturated flow beneath an ephemeral stream. *Water Resources Research* 34(9), 2137–2153.

Rühaak, W., Rath, V., Wolf, A., Clauser, C. (2008). 3D finite volume groundwater and heat transport modeling with non-orthogonal grids, using a coordinate transformation method. *Advances in Water Resources* 31(3), 513–524.

Russell, T.F., Heberton, C.I., Konikow, L.F., Hornberger, G.Z. (2003). A finite-volume ELLAM for three-dimensional solute-transport modelling. *Ground Water* 41(2), 258–272.

Sauty, J.P., Gringarten, A.C., Menjoz, A., Landel, P.A. (1982). Sensible energy storage in aquifers. 1. Theoretical study. *Water Resources Research* 18(2), 245–252.

Schmidt, T., Hellström, G. (2005). Ground source cooling—Working paper on usable tools and methods. EU Commission SAVE Programme and Nordic Energy Research.

Seo, H.S., Šimůnek, J., Poeter, E.P. (2007). Documentation of the HYDRUS package for MODFLOW-2000, the U.S. Geological Survey modular ground-water model, GWMI 2007-01. Int. Ground Water Modeling Ctr., Colorado School of Mines, Golden.

Sidiropoulos, E., Tzimopoulos, C. (1983). Sensitivity analysis of a coupled heat and mass transfer model in unsaturated porous media. *Journal of Hydrology* 64, 281–298.

Signorelli, S., Bassetti, S., Pahud, D., Kohl, T. (2007). Numerical evaluation of thermal response tests. *Geothermics* 36, 141–166.

Sophocleous, M. (1979). Analysis of water and heat flow in unsaturated-saturated porous media. *Water Resources Research* 15(5), 1195–1206.

Sun, Z.F., Carrington, C.G. (1995). A new numerical scheme for convective dominated heat transfer in a porous medium with strong temperature gradients. *Transport in Porous Media* 21, 101–122.

Tsang, C.F., Buschek, T., Doughty, C. (1981). Aquifer thermal energy storage: A numerical simulation of Auburn University field experiments. *Water Resources Research* 17(3), 647–658.

USGS (2012). 3D Finite-difference groundwater flow model MODFLOW. *United States Geological Service*, http://water.usgs.gov/software/lists/groundwater.

Vandenbohede, A., Lebbe, L. (2011). Heat transport in a coastal groundwater flow system near De Panne, Belgium. *Hydrogeology Journal* 19, 1225–1238.

Voss, C., Provost, A.M. (2010). SUTRA, A model for saturated-unsaturated, variable-density ground-water flow with solute or energy transport. Version 2.2, Water Resources Investigations Report 02-4231, USGS Reston, Virginia, USA.

Werner, D., Kley, W. (1977). Problems of heat storage in aquifers. *Journal of Hydrology* 34, 35–43.

Wiberg, N.-E. (1983). Heat storage in aquifers analyzed by the finite element method. *Ground Water* 21(2), 178–187.

Woods, K., Ortega, A. (2011). The thermal response of an infinite line of open-loop wells for ground coupled heat pump systems. *International Journal of Heat and Mass Transfer* 54, 5574–5587.

Wu, Y.S., Ahlers, C.F., Fraser, P., Simmons, A., Pruess, K. (1996). Software qualification of selected TOUGH2 modules. Rep. LBNL-39490, Lawrence Berkeley National Laboratory Berkeley, California.

Xue, Y., Xie, C., Li, Q. (1990). Aquifer thermal energy storage: A numerical simulation of filed experiments in China. *Water Resources Research* 26(10), 2365–2375.

Yeh, G.-T., Luxmoore, R.J. (1983). Modeling moisture and thermal transport in unsaturated porous media. *Journal of Hydrology* 64, 299–309.

Zheng, C. (1990). MT3D, A modular three-dimensional transport model for simulation of advection, dispersion, and chemical reactions of contaminants in groundwater systems. Report to the Kerr Environmental Research Laboratory, US Environmental Protection Agency, Ada, Oklahoma.

Zheng, C., Wang, P.P. (1999). MT3DMS: A modular three dimensional multi-species transport model for simulation of advection, dispersion and chemical reactions of contaminants in groundwater systems; Documentation and user's guide. U.S. Army Engineer Research and Development Center Contract Report SERDP-99-1, Vicksburg, Mississippi.

Chapter 5

Long-term operability and sustainability

The question of **sustainability** of thermal use of the shallow underground arises in various contexts. There are different definitions and, sometimes, subjective interpretations of sustainable use. Let us first take a technological perspective, which refers to the prolonged production ability. The issues have to be addressed separately for systems in relatively impermeable media and systems in prolific aquifers. In the latter case, one has to distinguish between closed and open systems.

The long-term viability of closed systems limits the amount of heat that can be abstracted sustainably, be it by one borehole heat exchanger (BHE) or a whole set of them. If that limit is overstepped, the temperature at the BHE drops below the freezing point of the working fluid and prevents the system from functioning. Already at prefreezing temperature, the efficiency of the system will of course drop considerably.

To assess the long-term development, thermal modeling is a useful tool. One would have to simulate the temperature distribution for a given heat abstraction and operation time, and check the temperature at the exchanger in order to judge the feasibility. The long-term viability depends on the presence of all users combined. The superposition of several users may lead to a nonsustainable heat abstraction, where a single user would still be doing fine. The question of "stealing" heat from others has to be treated in a heat management scheme for a formation. Regulations try to avoid interaction of neighboring users by defining minimum distances between competitive BHEs. In some countries, simply minimum distances to the property line are defined (Hähnlein et al. 2010a). However, by ignoring site-specific characteristics and potential groundwater flow, such constraints can promote but hardly optimize sustainable use.

5.1 SYSTEMS IN LOW PERMEABLE MEDIA

In the strictest sense, **sustainable use** means that a reservoir is exploited at a rate that does not lead to a decline in the resource in the future. Even though the stored energy in the upper hundred meters of the Earth's crust is vast, current technologies only allow uneven, local use. The permanent flux of geothermal heat in Switzerland is maximally about 100 mW m⁻². However, if one assumes that a single geothermal heat pump requires about 3 kW of heat power, this would mean that an area of at least 30,000 m² would be necessary per BHE. The exclusive use of this energy would allow only for relatively few users in concentrated settlements.

A single BHE, or a larger collection of BHEs, will therefore eventually empty the storage of utilizable heat, while at the same time lowering the temperature over a certain volume of the underground. On the other hand, the decreasing temperature at the BHE will create a gradient, also supplying heat from the surface, which can eventually contribute to a larger heat flux than the geothermal flux. For a sustainable operation in the long run, the system has to rely on the combined heat fluxes from the depth and from the surface. While the geothermal flux can be considered a constant flux boundary condition at depth, the surface can be considered a constant temperature boundary condition for all practical considerations.

The basic situation is shown in Figure 5.1: At steady state, the abstracted heat power at the BHE is provided by two contributions, the geothermal heat flux and the heat flux from the atmosphere. The BHE forms a "heat catchment" from which it draws the abstracted heat flux. Before steady state is reached, the heat flux at the BHE has an additional contribution stemming from cooling the surroundings of the BHE.

With increasing abstraction of heat at the BHE, the "heat catchment" of the exchanger is growing, while the temperature of the exchanger is decreasing. In order to guarantee a long lifetime of a BHE, that is, time

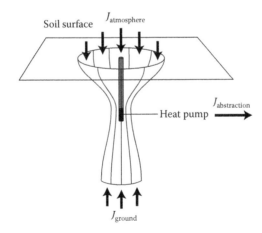

Figure 5.1 Heat catchment and heat balance around a BHE (schematic).

with temperatures at the BHE allowing an efficient production of heat, a proper design of the abstraction in connection with already existing abstractions is required.

The applied technologies cause **temperature anomalies** that may exist temporally, stabilize and arrive at steady state, or continuously change and evolve. The strict requirement of arriving at a viable steady state is sometimes relaxed. According to Rybach (2003), sustainability means the ability of the production system to sustain production levels over several decades to a couple of centuries. Similarly, Rybach and Eugster (2010) and Axelsson et al. (2001) state that for each geothermal system and for each mode of production, there exists a certain level of maximum energy production, below which it will be feasible to sustain constant energy production from the system for a very long time (100–300 years). The question of sustainability thus translates into the question of how long can a system operate without significant decline of production (Rybach and Mongillo 2006). As a more elaborate definition of sustainable operation, Hähnlein et al. (2013) suggest adopting the four modes defined by Axelsson (2010) for sustainable deep and/or high enthalpy geothermal utilization: (1) constant production on the sustainable level, where sustainability is related to the production ability of the system over an indefinitely long period; (2) stepwise increase in production until the sustainable level is achieved; (3) cyclic production (with an alternation of excessive production and periods of dormancy to allow for recovery); and (4) an excessive production followed by a reduced, steady production. Apparently, it is common to define a certain time frame, which serves as a premise based upon the assumption that sustainable use is reachable. In the best case, energy deficits are only temporary, of short duration, and, for instance, balanced during seasonal operation throughout each year. In the worst case, after the given time frame, the reservoir is depleted, and it will take a long time for replenishment if it is possible at all. In the ideal case, a viable steady state (or quasi–steady state with seasonal fluctuations) is established.

The **maximum heat flux** to a BHE in equilibrium can be determined by a numerical computation with a temperature boundary condition at the BHE set at the minimum allowable temperature, and applying the geothermal heat flux at the lower boundary, as well as the mean soil temperature at the upper boundary. Of course, the situation can be alleviated by adding heat in times where cooling of buildings rather than heating is required. In the climatic conditions of Switzerland, for example, the cooling requirements in large public buildings are only one third in terms of thermal energy required for heating. In the following, this situation is implicitly covered by considering the average net heat flux, which is the difference between heat abstraction and heat injection, as abstracted heat flux.

The use of the analytical solutions (without aquifer flow) discussed in Chapter 3 is feasible for crude estimates. First, a superposition with image sources, mirrored at the upper (fixed temperature) boundary, and the addition of the natural geothermal temperature gradient are necessary. As the analytical solutions have a flux boundary condition at the heat pump, that is, the abstraction rate of heat, the required temperature in the well has to be transformed to an equivalent heat abstraction rate iteratively (e.g., Hecht-Méndez et al. 2013). The singularity of analytical solutions for point or line sources is avoided by evaluating temperatures at the radius of the borehole (Beck et al. 2013).

Only a few **field studies** exist with long-term monitoring of temperatures. For a 105-m-long coaxial-tube BHE installed near Zurich (Switzerland), in **Elgg**, over 25 years of detailed observations gave insight into the thermal evolution of the ground (Eugster 2001; Rybach and Eugster 2002; Rybach and Eugster 2010; Figures 5.2 and 5.3). The BHE supplies a single-family house at SPF < 3 with peak thermal power of around 70 W m^{-1}. Temperature sensors (accuracy of 0.1 K) are installed at depths of 1, 2, 5, 10, 20, 35, 50, 65, 85, and 105 m at lateral distances of 0.5 and 1 m from the BHE. While at this location there is groundwater at a shallow depth of only a few meters, with a sequence of more or less productive aquifer layers, groundwater flow velocity was considered negligible and the case can be considered as an example for quasi-impermeable media. The temperature profiles recorded during the period of 1986–2001 clearly show the atmospheric influences down to a depth of about 15 m. Within the first few years, the thermal influence from seasonal energy extraction is most pronounced, with about 1–2 K smaller temperatures than in undisturbed ground. It is shown by continuous monitoring and supported by numerical modeling that at short distances from the BHE, the temperature decline continuously decelerates with small fluctuations within 0.5 K that are attributed primarily to the annually slightly varying heating demands (Figure 5.2). By simulation, it is revealed that the thermal anomaly evolves laterally at a very low rate and reaches several decameters within the operation period of 30 years. It is also shown that the heat sink promotes not only lateral heat flux to balance the artificial deficit around the BHE but also vertical heat flux from the atmosphere and geothermal flux from below. No steady state was reached in the 15 years of operation. Long-term simulation was used to examine the effect of abrupt BHE operation stop. Similar to the operation phase, initial thermal response of the ground right after the stop was most significant. Temperatures increased strongly during the first years, but to a lesser extent later. This reflects the effect of conduction, which decreases with thermal recovery as it is driven by the decreasing temperature gradient. Accordingly, about 30 years is needed to arrive at conditions with minor, but negligible, thermal anomalies. This means that about the same amount of time is needed as the system was operated to let it recover. In comparison, for deep geothermal energy use with Engineered Geothermal Systems (EGS), after a similar time of

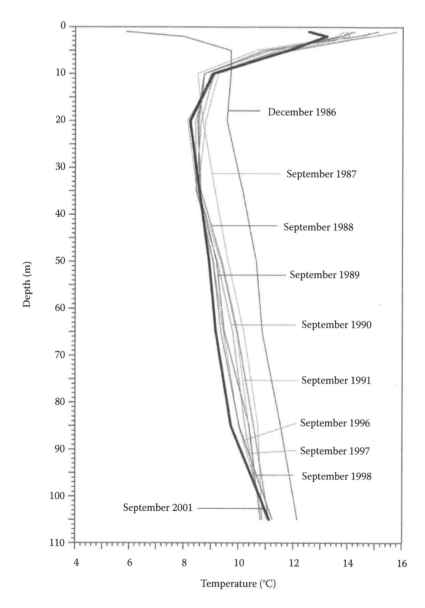

Figure 5.2 **(See color insert.)** Elgg site (Switzerland): measured ground temperature profiles at 0.5 m distance from a 105 m deep operating BHE, repeatedly measured over 15 years. (From Rybach, L. and Eugster, W.J., *Geothermics* 39, 365–369, 2010.)

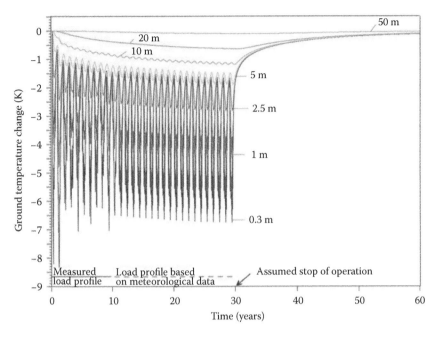

Figure 5.3 **(See color insert.)** Elgg site (Switzerland): simulated ground temperature changes of a BHE relative to the undisturbed situation in December 1986 over 30 years of operation and 30 years of recovery. (From Rybach, L. and Eugster, W.J., *Geothermics* 39, 365–369, 2010.)

operation, much higher recovery periods of up to 100 years (Tester et al. 2006) are needed as only the geothermal flux is available to "heal" the thermal anomaly.

5.2 THERMAL EVOLUTION IN AQUIFERS

There are many factors that influence the **evolution of thermal anomalies around geothermal applications in aquifers.** Among these are the technology type, such as open or closed systems, their size (single or multiple) and installation depth, the operation mode, and the hydraulic and thermal conditions in the subsurface. An additional role is played by the potential hydraulic and/or thermal influence of neighboring or upstream installations (Ferguson 2009). Especially in densely populated areas, competitive use of aquifers by upstream and downstream users is frequent and regulations ideally avoid interaction between adjacent systems. Minimum distances between different BHEs of neighboring installations in a range of 5–20 m are recommended (Hähnlein et al. 2010a).

For **closed systems in aquifers**, the situation in Figure 5.1 is changed, compared to the systems in relatively impermeable media, by additional heat fluxes provided by the advection of heat by the aquifer water flow and by groundwater recharge. On one hand, this increases the available heat flux. On the other hand, a new limitation is introduced by the environmental limits on the downstream temperature reduction, as discussed in the introduction. Now, two temperatures have to be checked for sustainability, the temperature T_b at the BHE (radius of borehole) and the temperature T_d at a given distance in the downstream. T_b should be above the minimum working temperature of the heat exchanger, and the difference between the original temperature and T_d should not be more than the given environmental limit at the location.

The existence of substantial groundwater flow is beneficial. As demonstrated by Hähnlein et al. (2010b), for a synthetic example with a closed system, high groundwater flow velocities, such as observed in gravel aquifers, lead to more elongated thermal plumes than smaller velocities in silty or sandy aquifers. It is also shown that under their assumptions, high groundwater flow velocity can yield thermal recovery within one year in seasonal use mode. This is promoted by combining heating and cooling operations, and for a specific site, recovery will depend on the local thermal and hydraulic conditions as well as the specific heat extraction and injection rates.

Using directly pumped **groundwater from wells** for heat extraction influences the hydraulic conditions in aquifers, and proper planning requires a hydrogeological analysis. For such open systems, which typically use multiple groundwater pumping and injection wells, thermal feedback between the wells is of major concern (Ferguson and Woodbury 2006). This may be avoided or mitigated by increasing the distance between wells or by general spatial rearrangement, by lowering the pumping rates, and/or by considering the viability of a balanced, seasonally reversible scheme (Banks 2008). The condition for sustainability in the open case is more easily applied. Now the legal limit on cooling applies (at most 3 K in a distance of 100 m downstream of the infiltration in Switzerland). This means that the maximum extractable heat flux J ($W = J\ s^{-1}$) can be estimated by the formula

$$J = \rho_w c_w Q \Delta T \tag{5.1}$$

where ΔT is the allowed temperature reduction of the reinfiltrated water and Q ($m^3\ s^{-1}$) is the infiltration rate. The application of analytical solutions for rough design purposes is more limited, as the modification of the flow connected to the reinfiltration does not allow for a closed solution of the temperature field. In a crude approximation, injections, which are not too close to each other, can be represented by line sources (transverse to flow) with a width corresponding to the asymptotic width of the injected water flow. This width can be determined analytically with the tools described in Chapter 3.2.1.

In general, numerical simulations have to be applied to inspect the thermal evolution and support optimal design. A challenge will be to capture transient hydraulic conditions of aquifers, such as dynamic changes in flow velocity or flow direction. This is particularly important in the vicinity of surface water bodies. Highly dynamic systems add complexity. At the same time, in comparison to sluggish conduction-dominated conditions, higher groundwater flow velocity means amplified heat or cold transfer by advection. As a consequence, a seasonal system can recover faster, a thermal anomaly can evolve less far into the downstream, and thermal conditions can more easily reach a steady state. As an example, for the Paris Basin, where doublet well systems have been successfully operated since the 1970s, no reduction in production temperature or substantial water level drawdowns has been reported (Rybach and Mongillo 2006). In contrast, Bonte et al. (2011) reported on a survey of 67 aquifer thermal energy systems (ATES) operated in the Netherlands that revealed that none of these systems was in thermal equilibrium. From the perspective of sustainable use, however, sufficient productivity may still be feasible. This is shown, for example, by Mégel and Rybach (2000) for a doublet system operated at a thermal capacity of 15 MW for district heating close to Basel. Numerical simulation revealed that an acceptable temperature decrease of 1.4 K will occur during the first 20 years. This rate substantially declines and much smaller values of 0.15 K in 10 years are expected after 300 years of operation.

5.3 FURTHER CRITERIA OF SUSTAINABILITY

Apart from criteria based on long-term performance of a shallow geothermal system and renewability of the thermal reservoir in the ground, sustainability is often also evaluated from environmental or economic points of view.

The **environmental aspects** concerning the temperature change have been discussed already in Chapter 1. However, the quality of the groundwater can also be influenced when the working fluid of a heat pump leaks out and consequently pollutes groundwater. For this reason, some regional agencies are reluctant to give any licenses for heat pumps in aquifers at all. The potential of polluting the downstream can be estimated by simulating the pollutant cloud forming in the downstream after complete working fluid loss. The potential depends on the type and quantity of cooling fluid used and can be kept small by regulating the allowable fluids.

Secondary effects on groundwater chemistry and biology are considered uncritical in most cases, when operating within the temperature limits of common guidelines (Tables 5.1 and 5.2). In open systems, wells are operated. The most critical (bio-)chemical consequence, which is often observed, is that clogging occurs in reinjection wells, and thus periodic back-flushing may be necessary to maintain long-term operability. In the aquifer, induced thermal gradients influence mixing processes. Furthermore, increase in temperature accelerates chemical reactions and modifies geochemical equilibria of minerals, oxygen saturation, and gas solubility. Through drilling of boreholes or installation of wells, new connections and flow paths in the subsurface can be created, and thus careful attention has to be given to potential effects, such as gypsum swelling. Likewise, cross-aquifer flow between contaminated and pristine aquifers needs to be avoided.

Thermal use of shallow aquifers may directly affect groundwater and groundwater-dependent ecosystems. In aquifers, microbes and groundwater invertebrates are important components of the ecosystem, and their

Table 5.1 Possible processes, effects, and their potential impact for shallow open geothermal energy systems

Process	Effects	Follow-up event	Potential impact	Significance
Temperature increase	Enhanced microbiological activity	Mineral precipitation	Clogging	++
			Biofilms	+
		Biofouling		−
		Slime production	Clogging	−
		Mass explosion		−−
		Sedimentation of iron ocher		++
		Corrosion		−
	Increase in mineral solubility (e.g., iron, manganese)[a]	Increase in mineral concentration in the groundwater (e.g., sedimentation of iron ocher)	Clogging	++
		Mass explosion (algae and bacterial growth)[a]		++
Temperature decrease	Increase in CO_2 solubility	Increased carbonate load	Clogging	++
Algae growth	Lowering pH Removing CO_2	Mineral precipitation[a]	Clogging	−
Shifting of material (solifluction)	Increase in holes		Changes in flow regime	−
	Accumulation of material			−
			Clogging	−

Source: After Hähnlein, S. et al. (2013), Sustainability and policy for the thermal use of shallow geothermal energy. *Energy Policy*, 59, 914–925.

[a] Critical iron concentration 40.1 mg/l and critical manganese concentration 40.05 mg/l.

Table 5.2 Physical, chemical, and biological consequences on open and closed systems and the thermally influenced area of ground-source heating pump systems (i.e., soil, ground, aquifer)

	Affected System			
	Open System		Closed System	
Follow-up event	System	TAA	System	TAA
Algae growth	+	+	−	+
Appearance of temperature anomalies	−	+	−	+
Changes in bacterial and faunal community		+		+
Changes in microbiological activity	+	+	−	+
Debonding	−	−	+	−
Gas solubility	−	++	−	+
Hydrological circuit/perforation of separating layers	+	−	+	−
Hydrological feedback	+		−	
Influence on surface ecosystem	−	+	−	++
Solifluction	−	+	−	−
Thermal feedback	+		−[a] / +[b]	

Source: After Hähnlein, S. et al. (2013), Sustainability and policy for the thermal use of shallow geothermal energy. *Energy Policy,* 59, 914–925.

Note: Follow-up event: (−) does not impact, (+) impact, (++) more pronounced impact on the system or temperature affected area (TAA).

[a] For single GSHP systems.
[b] For multiple GSHP systems.

diversity, composition, and functionality are influenced by the temperature. However, groundwater microbiology is a very young area of research, and general conclusions are hard to obtain. Hähnlein et al. (2013) reviewed four studies on the effect of temperature by shallow geothermal systems. It was found that there exists a tolerable range of temperature variations, which should be minimized in intensity, expansion, and duration. In the literature (Brielmann et al. 2009, 2011), we find acceptable ranges of ±6 K, and therefore, it is concluded that within the regulated temperature thresholds, impact of shallow geothermal energy use is only minor.

In addition to the direct site-specific consequences, **secondary environmental impacts** that are associated with the life cycle of a technology may be integrated in sustainability assessment. As pointed out by Saner et al. (2010), these are mainly controlled by the primary power consumption of the heat pump and thus the seasonal performance factor (SPF). The most common applications are closed systems, with far more than 1 million applications in Europe. An average European system that has been operating for

20 years would generate carbon dioxide (equivalent) emissions of 63 t at the present electricity mix (Saner et al. 2010). This is still less than traditional space heating technologies. Average savings are 35% in comparison to oil-fired boilers and 18% to gas-furnace heating systems. These figures refer to standard one-family houses. The use of shallow aquifers for combined heating and (passive) cooling, as it is common in bigger applications such as office buildings or district heating systems, is even more environmentally attractive.

Technical criteria and productivity can be quantitatively measured and predicted, and they have direct economic implications. In contrast, environmental or ecological criteria are often considered of secondary importance. They are sometimes difficult to evaluate, with distinct results depending on the assessment concept employed, the geological and groundwater conditions, the location of the system, and finally, the social environment involved (Hähnlein et al. 2013).

REFERENCES

Axelsson, G. (2010). Sustainable geothermal utilization—Case histories; definitions, research issues and modelling. *Geothermics* 39, 283–291.

Axelsson, G., Gudmundsson, A., Steingrimsson, B., Palmason, G., Armansson, H., Tulinius, H., Flovenz, O., Björnsson, S., Stefansson, V. (2001). Sustainable production of geothermal energy: Suggested definition. *IGA-News, Quarterly* 43, 1–2.

Banks, D. (2008). An Introduction to Thermogeology: Ground Source Heating and Cooling. Blackwell Publishing, Oxford, UK.

Beck, M., Bayer, P., de Paly, M., Hecht-Méndez, J., Zell, A. (2013). Geometric arrangement and operation mode adjustment in low-enthalpy geothermal borehole fields for heating. *Energy* 49, 434–443.

Bonte, M., Stuyfzand, P.J., Huismann, A., Van Beelen, P. (2011). Underground thermal energy storage: Environmental risks and policy developments in the Netherlands and European Union. *Ecology and Society* 16(1), 22.

Brielmann, H., Griebler, C., Schmidt, S.I., Michel, R., Lueders, T. (2009). Effects of thermal energy discharge on shallow groundwater ecosystems. *FEMS Microbiology Ecology* 68, 273–286.

Brielmann, H., Lueders, T., Schreglmann, K., Ferraro, F., Avramov, M., Hammerl, V., Blum, P., Bayer, P., Griebler, C. (2011). Oberflächennahe Geothermie und ihre potenziellen Auswirkungen auf Grundwasserökosysteme. *Grundwasser* 16, 77–91.

Eugster, W.J. (2001). *Langzeitverhalten der EWS-Anlage in Elgg (ZH)—Spotmessung im Herbst 2001.* Swiss Federal Office of Energy, Bern, Switzerland, 14 pp.

Ferguson, G. (2009). Unfinished business in geothermal energy. *Ground Water* 47(2), 167.

Ferguson, G., Woodbury, A.D. (2006). Observed thermal pollution and post-development simulations of low-temperature geothermal systems in Winnipeg, Canada. *Hydrogeology Journal* 14, 1206–1215.

Hähnlein, S., Bayer, P., Blum, P. (2010a). International legal status of the use of shallow geothermal energy. *Renewable and Sustainable Energy Reviews* 14, 2611–2625.

Hähnlein, S., Bayer, P., Ferguson, G., Blum, P. (2013). Sustainability and policy for the thermal use of shallow geothermal energy. *Energy Policy*, 59, 914–925.

Hähnlein, S., Molina-Giraldo, N., Blum, P., Bayer, P., Grathwohl, P. (2010b). Ausbreitung von Kältefahnen im Grundwasser bei Erdwärmesonden. *Grundwasser* 15, 123–133.

Hecht-Méndez, J., de Paly, M., Beck, M., Bayer, P. (2013). Optimization of energy extraction for vertical closed-loop geothermal systems considering groundwater flow. *Energy Conversion and Management* 66, 1–10.

Klotzbücher, T., Kappler, A., Straub, K.L., Haderlein, S.B. (2007). Biodegradability and groundwater pollutant potential of organic anti-freeze liquids used in borehole heat exchangers. *Geothermics* 36, 348–361.

Mégel, T., Rybach, L. (2000). Production capacity and sustainability of geothermal doublets. *Proceedings World Geothermal Congress 2000*, International Geothermal Association, Beppu-Morioka, Japan, pp. 849–854.

Rybach, L. (2003). Geothermal energy: Sustainability and the environment. *Geothermics* 32, 463–470.

Rybach, L., Eugster, W.J. (2002). Sustainability aspects of geothermal heat pumps. *Proceedings 27th Workshop on Geothermal Reservoir Engineering*, Stanford University, Stanford, California, pp. 50–64.

Rybach, L., Eugster, W.J. (2010). Sustainability aspects of geothermal heat pump operation, with experience from Switzerland. *Geothermics* 39, 365–369.

Rybach, L., Mongillo, M. (2006). Geothermal sustainability—A review with identified research needs. *Geothermal Resources Council (GRC) Transactions* 30, 1083–1090.

Saner, D., Juraske, R., Kübert, M., Blum, P., Hellweg, S., Bayer, P. (2010). Is it only CO_2 that matters? A life cycle perspective on shallow geothermal systems. *Renewable and Sustainable Energy Reviews* 14, 1798–1813.

Tester, J.W., Anderson, B.J., Batchelor, A.S., Blackwell, D.D., DiPippo, R., Drake, E.M., Garnish, J., Livesay, B., Moore, M.C., Nichols, K., Petty, S., Toksöz, M.N., Veatch, R.W. (2006). *The Future of Geothermal Energy—Impact of Enhanced Geothermal Systems (EGS) on the United States in the 21st Century*. Massachusetts Institute of Technology and US Department of Energy, Cambridge, MA, USA, 332 pp.

Chapter 6

Field methods

For the design and optimal performance of geothermal systems, various types of parameters such as **economical, technical, design, hydraulic, and thermal parameters,** have to be specified. For example, Blum et al. (2011), who studied the technoeconomic and spatial analysis of more than 1000 vertical ground-source heating pump (GSHP) systems with a heating demand of 11 ± 3 kW in southwestern Germany, concluded that subsurface characteristics are presently inadequately considered for the design of such GSHP systems. In this chapter, we merely discuss the most relevant thermal input parameters for the heat transport in the subsurface and design of geothermal systems using field methods such as **thermal response tests** (TRTs) and **thermal tracer tests** (TTTs). In the governing heat transport equations provided in Chapter 2, thermal diffusivities and hydraulic and thermal conductivities are important for heat transport simulations and design studies of closed and open geothermal systems. Hence, the focus is set on measurement techniques for determining these key hydraulic (K_w) and thermal parameters (λ_m, β_L, β_T). Values can be obtained both in the laboratory and in the field. The latter, being crucial for larger scale geothermal systems, is particularly considered here.

6.1 HYDROGEOLOGICAL FIELD METHODS

For the design of geothermal systems, the knowledge of subsurface characteristics is crucial, even more for open geothermal systems (e.g., Banks 2008). Depending on the type of investigation (e.g., water supply, contaminant, or heat transport), **standard hydrogeological field methods,** such as borehole flowmeter tests, slug tests, hydraulic pumping tests, and dye tracer tests, are typically used (e.g., Molz et al. 1989; Kruseman and De Ridder 1990; Fetter 2001; Schwartz and Zhang 2003). For example, for water supply and also for geothermal investigations of open systems, an integral evaluation of aquifer hydraulic conductivity by hydraulic pump tests is a standard approach. However, for optimal designs of aquifer thermal storage systems (ATES), more detailed knowledge of the spatial distribution of the hydraulic conductivity in the aquifer might be necessary. Hence, other hydraulic methods, such as **hydraulic tomography** and **direct-push methods**, are increasingly applied to describe the spatial distribution of the hydraulic conductivity at higher resolution than standard hydrogeological field methods do (e.g., Brauchler et al. 2003; Butler 2005; Illman et al. 2010; Lessoff et al. 2010). Despite the advantages of these evolving techniques, there are also limitations, the practicability of the hydraulic tomography in the field, and the very local insight obtained by direct-push methods, for example. A detailed discussion on the viability and application windows of these techniques, however, is beyond the scope of this book, and we refer to the study by Bohling and Butler (2010).

6.2 THERMAL RESPONSE TESTS

For the planning and design of large-scale GSHP systems, standard and enhanced **TRTs** are applied. The TRT is primarily used to estimate thermal properties of the subsurface and the heat transfer inside toward the tubes of the BHE. The principle of the TRT is similar to that of the standard hydraulic pumping tests (Raymond et al. 2011a), where an initially undisturbed system is perturbed and its response is subsequently monitored over time. Here, we will first provide an overview of the TRT, showing its development, setup, and application. Furthermore, we review analytical and numerical models for the evaluation of TRT. Finally, an analytical approach is discussed in more detail for the evaluation of groundwater-influenced TRT enabling the estimation of the local hydraulic conductivity.

6.2.1 Development of TRTs

The theoretical basis of TRT was originally developed by Choudhary (1976) and Morgensen (1983). The main idea of the TRT is to circulate a heat carrier fluid, such as water, in a BHE with a constant heating load and to continuously measure the temperature development of the fluid at the inlet and the outlet of the BHE. The first **mobile TRT device** called "TED" was developed in Sweden in 1995–1996 (Gehlin 1998; Figures 6.1 and 6.2). The test was then introduced and also improved in several other countries (e.g., Austin 1998; Austin et al. 2000; Gehlin 2002). In Germany, the first TRT with equipment based on the Swedish device was conducted in 1999 (Sanner et al. 2000). In the Netherlands, the TRT device was built with a reversible heat pump, supplying either warm or cold heat carrier fluid, and consequently could be used in both heating and cooling mode (Witte 2001). Worldwide, various types of such standard TRT equipment with different

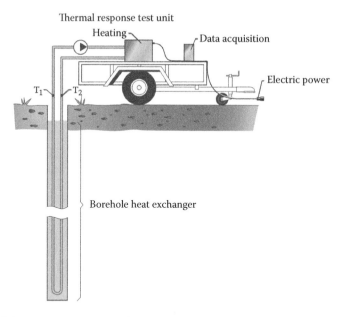

Figure 6.1 Schematic illustration of the setup for a standard TRT. T_{f1}: outlet fluid temperature; T_{f2}: inlet fluid temperature. (Illustrated by Claes-Göran Andersson. From Gehlin, S., Thermal response test method development and evaluation. Ph.D. thesis, Department of Environmental Engineering, Lulea University of Technology, Lulea, 2002.)

Figure 6.2 Mobile TRT device called "TED," which was developed in Sweden. (Photo by Signhild Gehlin. From Gehlin, S., Thermal response test method development and evaluation. Ph.D. thesis, Department of Environmental Engineering, Lulea University of Technology, Lulea, 2002.)

setups for heating and temperature monitoring are used (e.g., Roth et al. 2004; Sanner et al. 2005).

In addition to these standard TRT devices, various different types of so-called **enhanced TRT** were developed by various groups (e.g., Heidinger et al. 2004; Wagner and Rohner 2008; Raymond et al. 2010), where depth-dependent temperature series are measured to estimate depth-specific thermal properties. Heidinger et al. (2004) developed the so-called **enhanced geothermal response test** (EGRT), where both the heating and the temperature measurement, using an optical fiber sensor cable, are integrated in the borehole heat exchanger (BHE). Both systems are installed between the U-tubes of the BHE and the borehole wall inside the backfilling providing depth-specific thermal properties. Similarly, Wagner and Rohner (2008) measure the vertical temperature profile inside the U-tubes using a non-wired temperature probe, while Fuji et al. (2009) use optical fiber sensors. Both methods also provide insight into depth-specific thermal properties. In the following chapters, however, we focus only on the standard TRT.

6.2.2 Setup and application of TRTs

After completion of the borehole drilling, the borehole is typically equipped with a BHE consisting, for example, of double U-shaped polyethylene pipes, which in many countries are subsequently backfilled with a **cement–bentonite suspension**. The latter is left for several days until it is hardened and the released reaction heat has subsided. Before the TRT is started, the undisturbed ground temperature can be determined during an initial circulation phase without heating or cooling. The TRT is initialized by introducing a constant heating or cooling load, typically ranging between 30 and 80 W m^{-1}. The heating load during the test should be kept constant for the standard TRT evaluation. However, this is often challenging (Poppei et al. 2006). Furthermore, **external influences**, such as direct sunlight or seeping rainwater, can influence the apparatus temperature and therefore distort the test results. Hence, external pipes of the TRT devices should be comprehensively insulated (Figure 6.2), and the ambient air temperature should be also measured during the experiment to be able to assess its influence on the fluid temperature.

The total **test duration** can be as short as 12 to 20 h (Smith and Perry 1999), or 30 h recommended by Gehlin and Hellström (2003), and even longer periods of up to 50 h (Austin et al. 2000). Longer test periods, which tend to be more expensive, are desirable to average out diurnal variations. It is difficult to provide a universal recommendation. However, Beier and Smith (2003) provided a graphical method, which can be downloaded in the form of a spreadsheet (http://www.met.okstate.edu/FacultyandStaff/Beier/Beier_res.html), to determine minimum test duration based on subsurface and borehole properties. Based on a numerical model and for ideal conditions during the TRT, Signorelli et al. (2007) concluded that a duration of at least 50 h is required.

The TRT device generally includes a circulation pump connected with the pipes of the BHE and an electrical heater with a stable power supply. The flow rate is controlled to a constant value and monitored with two volumetric flow meters during the entire test duration. During the standard TRT, the following parameters are continuously measured and logged: heat carrier fluid flow rate, inflow and outflow heat carrier fluid temperatures, heat carrier temperature between circulation pump and heater, reference temperature in the trailer, and ambient air temperatures. The typical temperature data consist of curves showing the **ambient temperature and fluid temperature** development for both inlet and outlet over time (Figure 6.3). The data can finally be evaluated to determine subsurface properties, such as the **integral and effective thermal conductivity** λ_m of the subsurface and **thermal borehole resistance** R_{tb} of the BHE.

6.2.3 Evaluation of TRTs

The TRT can be generally evaluated using analytical and numerical models. The standard and also the enhanced TRT are most commonly evaluated using the **Kelvin line source theory**, which assumes an infinite, homogeneous, and isotropic medium and a constant and infinite heat source (Carslaw and Jaeger 1959). In addition, various alternative line source models (Chapter 3), for example, the **moving finite line source model** (Equation 3.60) and **cylindrical source models,** are also applied (Chapter 3.1.4). Besides those, a

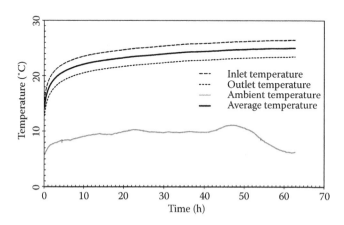

Figure 6.3 Example of the measured inlet and outlet fluid temperatures, the average fluid temperature in the pipes, and the ambient air temperature during a standard TRT in Germany.

variety of **numerical models** (Chapter 4) are also available for the analysis of TRT.

6.2.3.1 Analytical models

The standard TRT is most commonly evaluated using **Kelvin line source theory**. In the basic infinite line source model (Equation 3.10), a constant amount of energy is injected or extracted by conductive heat transport only. The temporal and spatial temperature changes around the line source can be determined and approximated as follows (Gehlin 2002; Signorelli et al. 2007; Wagner et al. 2013):

$$T(r,t) - T_0 = \frac{q_{tb}}{4\pi\lambda_m} \int\limits_{\frac{r^2}{4D_t t}}^{\infty} \frac{e^{-u}}{u} du = -\frac{q_{tb}}{4\pi\lambda_m} \text{Ei}\left[\frac{r^2}{4D_t t}\right] \approx \frac{q_{tb}}{4\pi\lambda_m}\left[\ln\left(\frac{4D_t t}{r^2}\right) - \gamma\right]$$

(6.1)

where T_0 is the initial or undisturbed temperature, q_{tb} (= J/H) is the heat flow rate per unit length of the borehole (W m^{-1}), λ_m is the effective thermal conductivity of the subsurface (W m^{-1} K^{-1}), and γ is the Euler constant (0.5772). If the time criterion $t \geq t_c \geq 5r_b^2 D_t^{-1}$ is fulfilled, the maximum error of the logarithmic approximation of the exponential integral is less than 10% (Witte et al. 2002). By increasing the time criterion, the latter can be decreased. For example, if $t_c \geq 20 r_b^2 D_t^{-1}$, the maximum error is only 2.5% (Wagner and Clauser 2005).

To determine the average heat carrier fluid temperature T_f, the thermal borehole resistance R_{tb} between the borehole wall and the circulating heat carrier fluid has to be considered, which is obtained by extension of Equation 6.1:

$$T_f - T_b = q_{tb} R_{tb}$$

(6.2)

$$T_f(t) = T_b(t) + q_{tb} R_{tb} = -\frac{q_{tb}}{4\pi\lambda_m} \text{Ei}\left[-\frac{r_b^2}{4D_t t}\right] + T_0 + R_{tb} q_{tb}$$

(6.3)

$$\approx \frac{q_{tb}}{4\pi\lambda_m} \ln(t) + q\left[R_{tb} + \frac{1}{4\pi\lambda_m}\left(\ln\left(\frac{4D_t t}{r_b^2}\right) - \gamma\right)\right] + T_0$$

where T_b is the temperature at the borehole wall (°C). To determine the effective thermal properties (λ_m and R_{tb}), two different approaches are generally feasible. The recorded TRT data can be either fitted by (1) a **linear**

regression (Gehlin 2002; Signorelli et al. 2007) or (2) a **parameter estimation technique** (Roth et al. 2004). The linear regression is based on the logarithmic approximation of Equation 6.3:

$$T_f = m \ln (t) + b \tag{6.4}$$

Hence, the slope m of Equation 6.4 is used to determine λ_m, and R_{tb} is estimated by the intercept with the y-axis b:

$$\lambda_m = \frac{q_{tb}}{4\pi m} = \frac{q_{tb}}{4\pi} \cdot \frac{\ln(t_2) - \ln(t_1)}{T_f(t_2) - T_f(t_1)} \tag{6.5}$$

$$R_{tb} = \frac{(b - T_0)}{q_{tb}} - \left(\frac{1}{4\pi\lambda_m} \left[\ln \left(\frac{4D_t t}{r_b^2} \right) - \gamma \right] \right) \tag{6.6}$$

This evaluation procedure was successfully applied in many different settings (Sanner et al. 2005). The main advantage of this variant is its simplicity, and for this reason, it has become the standard procedure. An example

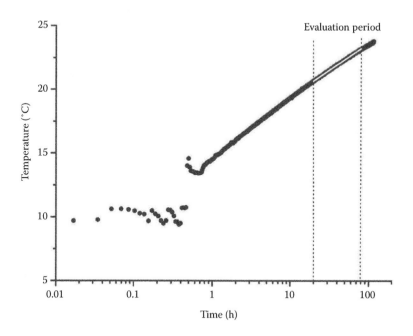

Figure 6.4 Duration time t versus average fluid temperature (T_f) of a standard TRT showing the linear regression for a selected evaluation period. (From Menberg, K. et al. *Grundwasser* 18, 103–116, 2013.)

is illustrated in Figure 6.4, showing the linear regression by plotting the time t along the x-axis in natural logarithmic scale (Menberg et al. 2013). Furthermore, an example for the parameter estimation technique is illustrated in Figure 6.5. It shows that depending on the selected procedure, also close-to-optimal solutions can be inspected in addition to the best fit. This example shows that given a certain tolerance for the fitting error, several value pairs of λ_m and R_{tb} are found. In practice, due to measurement errors, a tolerance range is often recommendable, and then the parameter estimation or a simple grid search is preferable to linear regression. However, the TRT may not deliver specific parameter values but rather correlated parameter pairs.

Using the temporal **superposition principle,** Raymond et al. (2011a) developed a TRT evaluation, which can also consider variable heat injection rates and can therefore also analyze the temperature recovery of a TRT by automatic optimization of the parameters using the solver function in Microsoft Excel. The spreadsheet can be downloaded as supporting information from the review paper by Raymond et al. (2011a). An example for the TRT analysis using the superposition principle, which is also discussed in detail in the review paper, is illustrated in Figure 6.6.

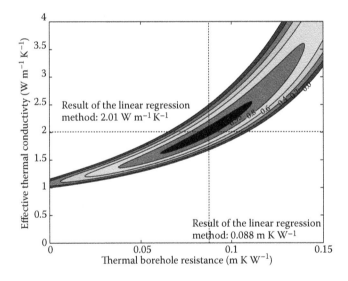

Figure 6.5 **(See color insert.)** Example of the parameter estimation technique for the evaluation of the thermal borehole resistance and the effective thermal conductivities showing the results of the model efficiencies (EF values) according to Loague and Green (1991). The results of the linear regression method are also shown. (From Wagner, V. Analysis of thermal response tests using advanced analytical and high resolution numerical simulations, Diploma thesis, University of Tubingen, Tubingen, Germany, 88 pp. 2010.)

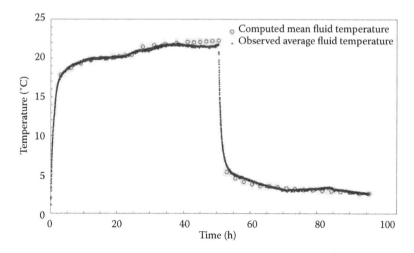

Figure 6.6 TRT data and analysis from the Doyen Mine in Québec (Canada) using the superposition principle and best fit for the entire test duration including the temperature recovery phase. (Modified after Raymond et al. 2011a.)

Due to the many assumptions for standard TRT evaluation using the Kelvin line source model, the **main limitations** arising are as follows:

1. It is impossible to evaluate the **first hours** of a TRT for the characterization of the subsurface. The reason is that the response signal is initially only influenced by the thermal properties of the BHE (Poppei et al. 2006); hence, the resulting values would be dominated by the BHE properties.

2. The assumption of a **constant heat injection** is often difficult to assure in practice. Several studies reported fluctuations for the heat input during the test duration (e.g., Eklöf and Gehlin 1996; Poppei et al. 2006; Witte et al. 2002). To overcome this issue, the TRT analysis suggested by Raymond et al. (2011a), which is able to consider variable heat injection rates, could be used.

3. A limitation of the standard TRT evaluation is the assumption of a **homogeneous undisturbed soil temperature.** For example, the geothermal gradient is not included in the standard TRT analysis (Witte et al. 2002). The influence of the geothermal gradient was comprehensively studied by Wagner et al. (2012) using a numerical model. The study shows that typical geothermal gradients (0°C to 5.2°C per 100 m) result in an underestimation of λ_m and R_{tb} using the Kelvin line source model. The estimation error may even exceed 10% for gradients of 5.2°C per 100 m.

4. The assumption that the studied medium is **homogeneous, isotropic, and infinite** is always doubtful, as BHEs often penetrate several different geological layers with different thermal and hydraulic properties. Raymond et al. (2011a) revealed that geological heterogeneity such as layering can result in an overestimation of λ_m. Alternatively, enhanced TRT or numerical models could be used to study the vertical distribution and heterogeneity of thermal properties (e.g., Raymond and Lamarche 2013).

5. **Horizontal groundwater flow** is also not considered by the conductive line source model. Yet, many studies have demonstrated the influence of increasing groundwater velocities on the estimation of the effective thermal conductivity (e.g., Bardenhagen et al. 2010; Witte et al. 2002). Signorelli et al. (2007) showed a significant influence of horizontal groundwater flow velocities higher than 0.1 m day^{-1} on the results of a TRT.

6. **Vertical groundwater flow** may also influence the results of a TRT. Borehole convection inside a BHE mainly appears in open boreholes, poorly grouted BHEs, or BHEs that are grouted with sand (Sanner et al. 2005).

In addition, a comprehensive **error analysis** of TRT, which was performed by Witte (2013), showed that measurement and theoretical errors such as parameter and model errors are about 5% for the thermal conductivity and 10%–15% for the borehole thermal resistance.

To overcome all restrictions and limitations such as variable heat injection rates, heterogeneities, and groundwater flow, enhanced TRT (e.g., Heidinger et al. 2004), the analytical approach by Wagner et al. (2013), or improved evaluation strategies like the finite and moving line source models (e.g., Molina-Giraldo et al. 2011), the superposition model by Raymond et al. (2011a), or numerical models (e.g., Signorelli et al. 2007) might be applied.

6.2.3.2 Numerical models

Numerical models have become increasingly popular, because they are able to account for spatial and temporal aspects that are typically ignored or not considered by analytical models, such as groundwater flow (Signorelli et al. 2007), specific borehole geometries, and heterogeneities of the hydraulic and thermal properties of the subsurface and the BHE. However, numerical models often need a large amount of data and information to demonstrate their advantage compared to analytical solutions. They are time-consuming and not justified for conventional TRT evaluations and cost-based geothermal projects. Nevertheless, various numerical models have been developed for BHE simulation and applied to TRT interpretation with parameter estimation techniques (e.g., Eskilson 1986; Diersch et al. 2011a,b; Raymond et al. 2011b; Wagner et al. 2012).

Eskilson (1986) developed the **superposition borehole model** (SBM), a FORTRAN-based code that is able to simulate the three-dimensional (3D) temperature field of one or several BHEs. In 1996, the SBM was integrated into the commercial transient energy simulation software package TRNSYS, the combination being called TRNSBM. Witte and van Gelder (2006) combined the latter with a parameter estimation procedure using the generic optimization package GenOpt. In addition, they performed two TRTs with and without controlled horizontal groundwater flow, where groundwater was pumped with a flow rate of 2.9 m^3 h^{-1} from an extraction well at a distance of 5 m from the studied grouted BHE. Without groundwater flow, λ_m was estimated to be 2.34 W m^{-1} K^{-1} and with groundwater flow 3.22 W m^{-1} K^{-1}, clearly demonstrating the influence of groundwater flow on the evaluation of λ_m. The simulation performed with TRNSBM showed that even for a small Darcy velocity of <3.5 m per year, the estimated λ_m would be 6% higher in comparison to purely conductive conditions.

Shonder and Beck (1999) developed a **one-dimensional** (1D) **finite difference** (FD) **BHE model**, which is also based on a parameter estimation technique for determining thermal properties from short-period TRT. They simulated the inlet and outlet temperatures and flows using a cylinder source model. They showed that the model is even accurate for short times, and therefore, early-time data from the experiment can be used, which is an advantage compared to the analytical cylinder source method. Gehlin (2002) also developed an explicit 1D FD BHE model, which consists of 18 cells coarsening in the radial direction from the center of the BHE. The first and second cells represent the heat carrier fluid and the grouting material, and the remaining cells represent the subsurface. The results of this 1D numerical model showed slightly higher values for the thermal conductivity and R_{tb} in comparison with the analytical line source model.

A transient **two-dimensional** (2D) **finite volume** (FV) **model** of a vertical BHE was developed by Yavuzturk and Spitler (1999). The 2D model also uses a parameter estimation algorithm by varying R_{tb} and thermal conductivities from grout and subsurface. Wagner and Clauser (2005) developed a parameter estimation technique using the **3D FD code** SHEMAT (Chapter 4). With the developed approach, it was also possible to estimate an integral heat capacity of the ground. They showed for the TRT analysis that the average variation of the heat capacity of around 20% may only cause a 2% difference in geothermal energy yield.

Signorelli et al. (2007) used the **3D finite element** (FE) **code** FRACTure for the TRT analysis, which was previously successfully applied for the simulation of a deep BHE in Switzerland (Kohl et al. 2002). Other existing FE numerical flow and heat transport codes (see Chapter 4), such as HydroGeoSphere (Raymond et al. 2011b) and FEFLOW (Diersch et al. 2011a), were extended to also simulate BHEs. In FEFLOW, the numerical strategy developed by Al-Khoury et al. (2005) was extended, adopted,

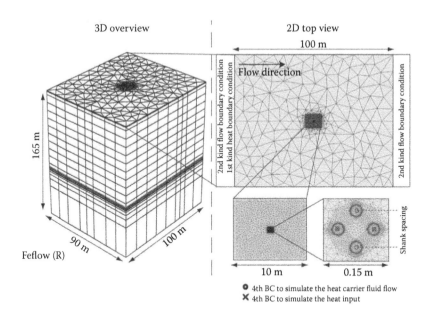

Figure 6.7 Left: 3D overview of the model domain and applied discretization. Right: 2D top view of the model domain and boundary conditions used. (Modified after Wagner, V. et al. *Renewable Energy* 41, 245–253, 2012.)

verified against an analytical solution, and applied for borehole thermal energy storage (BTES) consisting of 80 BHEs (Diersch et al. 2011b). Using FEFLOW (version 6.0), Wagner et al. (2012) studied the effects of (1) the in situ position of the U-shaped pipes of the BHE (shank spacing), (2) different geothermal gradients (i.e., nonuniform initial thermal distributions), and (3) thermal dispersivity (Figure 6.7). The results showed that the shank spacing and typical geothermal gradients have only minor effects (<10%) on the evaluation of λ_m and R_{tb}. However, given a constant groundwater flow velocity, varying thermal dispersivity values can have a significant impact on the evaluation of R_{tb}.

6.3 THERMAL TRACER TEST

When we use heat as a tracer (e.g., Anderson 2005; Saar 2011), we distinguish between long-term and short-term experiments (Wagner et al. 2014). **Long-term injection-storage experiments** are conducted to assess the performance of ATES (e.g., Molz et al. 1978; Sauty et al. 1982a,b; Xue et al. 1990; Palmer et al. 1992; Wu et al. 2008) and are also comprehensively discussed in Chapter 7. Such experiments are typically conducted with large volume injections of hot water (thousands of cubic meters) and with long-term monitoring of aquifer temperature changes (months to years). The main purpose of such large-scale field experiments is to assess the warm water storage capacity and/or recovery efficiencies of ATES. In addition, short-term active **TTTs** are infrequently conducted to derive hydraulic and thermal parameters (e.g., Shook 2001; Ma et al. 2012; Wagner et al. 2014). Here, for short periods, heated or cooled water is injected as a tracer, and then temperature changes are continuously measured in nearby observation and/or extraction wells. In some "heat tracer" experiments, thermal effects are induced by in situ heating of specific devices (e.g., Leaf et al. 2012), and one may categorize these applications as TRT-type variants without mass exchange.

In Table 6.1, selected short-term TTTs are summarized and discussed below with respect to configuration, hydrogeological setting, and test duration (Wagner et al. 2014). Keys and Brown (1978) conducted TTT in Texas, USA, by performing three **recharge tests** with various injection temperatures, water volumes, and rates. The injected water was supplied from a nearby playa lake with diurnal fluctuations of water temperatures between 13°C and 23°C. Up to 46 m away from the recharge well, the water was continuously monitored in five observation wells. The thermal pulses recorded in the wells were analyzed, and the hydraulic conductivity and its distribution were determined using laboratory and field data. Macfarlane et al. (2002) conducted a forced gradient injection test in a fractured porous sandstone aquifer in Kansas, USA. Heated water with 73°C was injected, and the temperature was monitored using distributed optical-fiber temperature sensing (DTS) in a pumped well at a distance of 13.2 m. The groundwater flow velocity derived, using the arrival time of the plume, is 2.7×10^{-5} m s^{-1}, which is 30 times larger than the estimated regional flow velocity. Leaf et al. (2012) performed three **open-well thermal dilution tests** in a fractured porous sandstone aquifer in Wisconsin, USA, using DTS also for the temperature monitoring. However, no flow velocities or hydraulic conductivities could be determined with the tests, which only provided information on the borehole flow regimes.

Vandenbohede et al. (2008a,b, 2009) conducted two single-well **push–pull tests** (PPTs) in a deep aquifer in Belgium. These tests were designed to evaluate the performance of a planned ATES. The obtained data were also exploited to examine the differences between solute and heat transport. They used an injection temperature for both PPT with 11.5°C, which was slightly colder than the ambient aquifer temperature of 15.8°C (Table 6.2). The 2D FD numerical model ReacTrans was adopted to simulate the field tests, in which the simulated solute (chloride) and heat transport were compared. They concluded that the most sensitive parameters are solute longitudinal dispersivity for the solute transport and thermal diffusivity for the heat transport (Vandenbohede et al. 2009). Ma et al. (2012) applied the 3D numerical groundwater flow model MODFLOW and MT3DMS/SEAWAT for studying

Table 6.1 Overview of short-term (<12 days) TTTs reported in literature

Location	Aquifer type	Injected volume (m³)	Injection rate (m³ h⁻¹)	Temperature difference (K)	Injection time (h)	Duration (days)	Observation wells	Remarks	Reference
Stewart site, Texas, USA	Unconsolidated porous aquifer	32,832	3283	−2.3 to + 7.7	240	10	5	Natural gradient test; variable injection temperature	Keys and Brown (1978)
Kansas, USA	Porous fractured sandstone aquifer	359.6	2.5	+55	173	7.2	1	Forced gradient test, one production and one observation well	Macfarlane et al. (2002)
Coastal plane, Belgium	Deep fine sand confined porous aquifer	188	3.9	−4.3	48.15	9.2	–	Only short-term push and pull test	Vandenbohede et al. (2008a,b, 2009)
Hanford site, USA	Unconfined and unconsolidated porous aquifer	156	16.3	−7.8	9.75	11.8	28	Parallel solute and TTT	Ma et al. (2012)
Wisconsin, USA	Porous fractured sandstone aquifer	Not specified	0.6–0.8	+2–7	2.6–3	0.2–0.3	–	Three open-well thermal dilution tests; wells intersect several aquifers	Leaf et al. (2012)
Lauswiesen site, Germany	Unconfined shallow porous aquifer	16	1.0	+11	8.0	4	5	Natural gradient test	Wagner et al. (2013)

Source: Wagner et al. Thermal tracer testing in a heterogeneous sedimentary aquifer: Field experiment and numerical simulation. Hydrogeology Journal. 2014.

Table 6.2 Examples of thermal dispersivity values reported in literature obtained by field experiments and synthetic modeling studies

Reference	Location	Method	Thermal dispersivity	Remarks
Andrews and Anderson (1979)	Cooling lake of a power plant in Wisconsin (USA)	FE model, trial and error adjustment procedure; seepage from the cooling lake	β_L = 0.1 m (best fit) β_L = 0.025 m	Longitudinal to transverse dispersivity ratio = 4; scale of the 2D model = 200 m
Sauty et al. (1982b)	Bonnaud (Jura, France); confined aquifer with 2.5 m thickness	Large volume injections (up to 1680 m³); calibrated using two 3D numerical models	β_L = 1 m (best fit)	Thermally influenced radius of about 13 m
Smith and Chapman (1983)	–	Synthetic study of a sedimentary basin (40 km wide and 5 km deep with a 1 km relief)	β_L = 100 m β_T = 10 m	2D FE flow and transport model
Xue et al. (1990)	Shanghai (China)	Single-, double-, and multiple-well experiments using flow rates of about 700 m³ day⁻¹	β_L = 3.30 m (best fit)	3D FD flow and transport model
Molson et al. (1992)	Borden aquifer (Canada)	Warm water injection (35°C) with volume of about 54 m³	β_L = 0.1 m β_{Th} = 0.01 m β_{Tv} = 0.005 m	3D FE flow and transport model
Su et al. (2004)	Russian River, California (USA)	Simulations of stream and groundwater temperature profiles for the six wells	β_L = 0.5 m β_T = 0.05 m	2D FD flow and transport model, best fit only with β_L
Ma et al. (2012)	Hanford site, Washington State (USA)	Combined bromide and heat tracer experiment with an injected volume of about 154 m³	β_L = 1.0 m β_{Th} = 0.1 m β_{Tv} = 0.01 m	3D FD flow and transport model, only β_L was used for calibration
Gelhar et al. (1992)	–	1D radial flow solution 2D numerical model 3D numerical model	β_L = 1.5 m β_L = 1.0 m β_L = 0.76 m	Review of data of field-scale dispersion in aquifers

Note: β_{Th}, transversal horizontal; β_{Tv}, transversal vertical.

heat transport at the MADE site. The comparison between the two heat transport models was used to investigate the influence of variable densities and viscosities (Table 6.2). They demonstrated that when the maximum temperature difference is within 15°C, the assumption of constant fluid density and viscosity has only negligible effects on the simulated temperature distribution (Ma and Zheng 2010). Wagner et al. (2014) conducted a TTT for the characterization of a shallow heterogeneous aquifer close to Tübingen, Germany (Table 6.2). A FEFLOW based 3D model was set up to reproduce the thermal anomaly observed after 8 hours of warm water injection.

These TTTs successfully demonstrated that aquifer structures and/or properties could be comprehensively studied by monitoring groundwater temperatures. Both long- and short-term experiments can particularly be used to estimate **thermal dispersivity** values, which are only sparsely reported in the literature (e.g., Molina-Giraldo et al. 2011). Examples of reported values, which range between 0.1 and 100 m depending on the scale of observation, are provided in Table 6.2.

REFERENCES

Al-Khoury, R., Bonnier, P., Brinkgreve, R. (2005). Efficient finite element formulation for geothermal heating systems. Part I: Steady state. *International Journal of Numerical Methods in Engineering* 63(7), 988–1013.

Anderson, M.P. (2005). Heat as a groundwater tracer. *Ground Water* 43(6), 951–968.

Andrews, C.B., Anderson, M.P. (1979). Thermal alteration of groundwater caused by seepage from a cooling lake. *Water Resources Research* 15(3), 595–602.

Austin, A.W. (1998). Development of an in situ system for measuring ground thermal properties. Master-Thesis, Oklahoma State University, Stillwater, Oklahoma.

Austin, A.W., Yavuzturk, C., Spitler, J. (2000). Toward optimum sizing of heat exchangers for ground-source heat pump systems-development of an in-situ system and analysis procedure for measuring ground thermal. *ASHRAE Transactions* 106(1), 831–842.

Banks, D. (2008). *An Introduction to Thermogeology: Ground Source Heating and Cooling*. Blackwell Publishing, Oxford, UK.

Bardenhagen, I., Brenner, K., Oberle, A. (2010). Der Einfluss von fließendem Grundwasser auf Thermal-Response-Tests. *bbr-Fachmagazin für Brunnen und Leitungsbau* 61, Sonderheft Geothermie, 34–41.

Beier, R.A., Smith, M.D. (2003). Minimum duration of in-situ tests on vertical boreholes. ASHRAE *Transactions* 109(3), 475–486.

Blum, P., Campillo, G., Kölbel, T. (2011). Techno-economic and spatial analysis of vertical ground source heat pump systems in Germany. *Energy* 36, 3002–3011.

Bohling, G.C., Butler, J.J. (2010). Inherent limitations of hydraulic tomography. *Ground Water* 48, 809–824.

Brauchler, R., Liedl, R., Dietrich, P. (2003). A travel time based hydraulic tomographical approach. *Water Resources Research* 39(12), 1370, doi:10.1029/2003WR002262.

Butler, J.J., Jr. (2005). Hydrogeological methods for estimation of hydraulic conductivity. In: Y. Rubin, S. Hubbard (Eds.), *Hydrogeophysics*. Springer, the Netherlands, pp. 23–58.

Carslaw, H.S., Jaeger, J.C. (1959). *Conduction of Heat in Solids*. 2nd ed., Oxford University Press, Oxford, UK.

Choudhary, A. (1976). An approach to determine the thermal conductivity and diffusivity of a rock in situ. Ph.D. Thesis, Oklahoma State University.

Diersch, H.-J.G., Bauer, D., Heidemann, W., Rühaak, W., Schätzl, P. (2011a). Finite element modeling of borehole heat exchanger systems. Part 1. Fundamentals. *Computers and Geosciences* 37, 1122–1135.

Diersch, H.-J.G., Bauer, D., Heidemann, W., Rühaak, W., Schätzl, P. (2011b). Finite element modeling of borehole heat exchanger systems. Part 2. Numerical simulations. *Computers and Geosciences* 37(8), 1136–1147.

Eklöf, C., Gehlin, S. (1996). TED-a mobile equipment for thermal response test. Diploma Thesis, Luleå Tekniska Universitetet, Sweden.

Eskilson, P. (1986). *Superposition Borehole Model, Manual for Computer Code*. Department of Mathematical Physics, University of Lund, Lund, Sweden.

Fetter, C.W. (2001). *Applied Hydrogeology*. Prentice Hall, New Jersey, 598 pp.

Fujii, H., Okubo, H., Nishi, K., Itoi, R., Ohyama, K., Shibata, K. (2009). An improved thermal response test for U-tube ground heat exchanger based on optical fiber thermometers. *Geothermics* 38(4), 399–406.

Gehlin, S. (1998). Thermal response test—In situ measurements of thermal properties in hard rock. Licentiate Thesis, Department of Environmental Engineering, Lulea University of Technology, Sweden.

Gehlin, S. (2002). Thermal response test method development and evaluation. Ph.D. thesis, Department of Environmental Engineering, Lulea University of Technology, Lulea.

Gehlin, S., Hellström, G. (2003). Influence on thermal response test by groundwater flow in vertical fractures in hard rock. *Renewable Energy* 28, 2221–2238.

Gelhar, L.W., Welty, C., Rehfeldt, K.R. (1992). A critical review of data on field-scale dispersion in aquifers. *Water Resources Research* 28(7), 1955–1974.

Heidinger, P., Dornstädter, J., Fabricius, A., Welter, M., Wahl, G., Zurek, K. (2004). EGRT—Enhanced Geothermal Response Test. 8. Geothermische Fachtagung, Landau in der Pfalz, 10–12, November 2004, pp. 319–325.

Illman, W.A., Zhu, J., Craig, A.J., Yin, D. (2010). Comparison of aquifer characterization approaches through steady state groundwater model validation: A controlled laboratory sandbox study. *Water Resources Research* 46, W04502, doi:10.1029/2009WR007745.

Keys, W.S., Brown, R.F. (1978). The use of temperature logs to trace the movement of injected water. *Ground Water* 16(1), 32–48.

Kohl, T., Brenni, R., Eugster, W.J. (2002). System performance of a deep borehole heat exchanger. *Geothermics* 31, 687–708.

Kruseman, G.P., de Ridder, N.A. (1990). *Analysis and Evaluation of Pumping Test Data*. International Institute for Land Reclamation and Improvement, Wageningen, the Netherlands. ISBN 90-70754-20-7.

Leaf, A.T., Hart, D.J., Bahr, J.M. (2012). Active thermal tracer tests for improved hydrostratigraphic characterization. *Ground Water* 50(5), 726–735.

Lessoff, S.C., Schneidewind, U., Leven, C., Blum, P., Dietrich, P., Dagan, G. (2010). Spatial characterization of the hydraulic conductivity using direct-push injection logging. *Water Resources Research* 46, W12502.

Loague, K., Green, R.E. (1991). Statistical and graphical methods for evaluating solute transport models: Overview and application. *Journal of Contaminant Hydrology* 7(1–2), 51–73.

Ma, R., Zheng, C. (2010). Effects of density and viscosity in modeling heat as a groundwater tracer. *Ground Water* 48(3), 380–389.

Ma, R., Zheng, C., Zachara, J.M., Tonkin, M. (2012). Utility of bromide and heat tracers for aquifer characterization affected by highly transient flow conditions. *Water Resources Research* 48(8), 1–18.

Macfarlane, A., Förster, A., Merriam, D., Schrötter, J., Healey, J. (2002). Monitoring artificially stimulated fluid movement in the Cretaceous Dakota aquifer, western Kansas. *Hydrogeology Journal* 10(6), 662–673.

Menberg, K., Steger, H., Zorn, R., Reuss, M., Proell, M., Bayer, P., Blum, P. (2013). Bestimmung der Wärmeleitfähigkeit im Untergrund durch Labor- und Feldversuche und anhand theoretischer Modelle. *Grundwasser*, 18, 103–116.

Molina-Giraldo, N., Bayer, P., Blum, P. (2011). Evaluating the influence of mechanical thermal dispersion on temperature plumes from geothermal systems using analytical solutions. *International Journal of Thermal Sciences* 50, 1223–1231.

Molson, J.W., Frind, E.O., Palmer, C.D. (1992). Thermal energy storage in an unconfined aquifer: 2. Model development, validation, and application. *Water Resources Research* 28(10), 2857–2867.

Molz, F.J., Morin, R.H., Hess, A.E., Melville, J.G., Guven, O. (1989). The impeller meter for measuring aquifer permeability variations: Evaluation and comparison with other tests. *Water Resources Research* 25(7), 1677–1683.

Molz, F.J., Warman, J.C., Jones, T.E. (1978). Aquifer storage of heated water: Part I—A field experiment. *Ground Water* 16, 234–241.

Morgensen, P. (1983). Fluid to duct wall heat transfer in duct system heat storage. *Proceedings of the International Conference on Surface Heat Storage in Theory and Practice*, Stockholm, Sweden, pp. 652–657.

Palmer, C.D., Blowes, D.W., Frind, E.O., Molson, J.W. (1992). Thermal energy storage in an unconfined aquifer: 1. Field Injection Experiment. *Water Resources Research* 28(10), 2845–2856.

Poppei, J., Schwarz, R., Mattsson, N., Laloui, L., Wagner, R., Rohner, E. (2006). Innovative improvements of thermal response tests. Tech. Rep., Swiss Federal Office of Energy.

Raymond, J., Lamarche, L. (2013). Simulation of thermal response tests in a layered subsurface. *Applied Energy* 109, 293–301.

Raymond, J., Robert, G., Therrien, R., Gosselin, L. (2010). A novel thermal response test using heating cables. *Proceedings of the World Geothermal Congress*, Bali, Indonesia, pp. 1–8.

Raymond, J., Therrien, R., Gosselin, L., Lefebvre, R. (2011a). A review of thermal response test analysis using pumping test concepts. *Ground Water* 49(6), 932–945.

Raymond, J., Therrien, R., Gosselin, L., Lefebvre, R. (2011b). Numerical analysis of thermal response tests with a groundwater flow and heat transfer model. *Renewable Energy* 36(1), 315–324.

Roth, P., Georgiev, A., Busso, A., Barraza, E. (2004). First in situ determination of ground and borehole thermal properties in Latin America. *Renewable Energy* 29(12), 1947–1963.

Saar, M.O. (2011). Review: Geothermal heat as a tracer of large-scale groundwater flow and as a means to determine permeability fields. *Hydrogeology Journal* 19(1), 31–52.

Sanner, B., Hellström, G., Spitler, J., Gehlin, S. (2005). Thermal response test—Current status and world wide application. *World Geothermal Congress*, Antalya, Turkey, pp. 1436–1445.

Sanner, B., Reuss, M., Mands, E. (2000). Thermal response test—Experience in Germany. *Proceedings of the 8th International Conference on Thermal Energy Storage, Terrastock 2000*, Stuttgart, Germany, pp. 177–182.

Sauty, J.P., Gringarten, A.C., Fabris, H., Thiery, D., Menjoz, A., Landel, P.A. (1982b). Sensible energy storage in aquifers—2. Field experiments and comparison with theoretical results. *Water Resources Research* 18(2), 253–265.

Sauty, J.P., Gringarten, A.C., Menjoz, A., Landel, P.A. (1982a). Sensible energy storage in aquifers—1. Theoretical study. *Water Resources Research* 18(2), 245–252.

Schwartz, F.W. Zhang, H. (2003). *Fundamentals of Ground Water*. 583 pages, John Wiley & Sons, New York, USA.

Shonder, J.A., Beck, J.V. (1999). Determining effective soil formation properties from field data using a parameter estimation technique. *ASHRAE Transactions* 105, 458–466.

Shook, G.M. (2001). Predicting thermal breakthrough in heterogeneous media from tracer tests. *Geothermics* 30(6), 573–589.

Signorelli, S., Bassetti, S., Pahud, D., Kohl, T. (2007). Numerical evaluation of thermal response tests. *Geothermics* 36, 141–166.

Smith, L., Chapman, D.S. (1983). On the thermal effects of groundwater flow—1. Regional scale systems. *Journal of Geophysical Research* 88(B1), 593–608.

Smith, M., Perry, R. (1999). In situ testing and thermal conductivity testing. *Proceedings of the 1999 GeoExchange Technical Conference and Expo*, Oklahoma State University, Stillwater, Oklahoma.

Su, G.W., Jasperse, J., Seymour, D., Constantz, J. (2004). Estimation of hydraulic conductivity in an alluvial system using temperatures. *Ground Water* 42, 890–901.

Vandenbohede, A., Louwyck, A., Lebbe, L. (2008a). Identification and reliability of microbial aerobic respiration and denitrification kinetics using a single-well push-pull field test. *Journal of Contaminant Hydrology* 95(1–2), 42–56.

Vandenbohede, A., Louwyck, A., Lebbe, L. (2009). Conservative solute versus heat transport in porous media during push-pull tests. *Transport Porous Media* 76(2), 265–287.

Vandenbohede, A., Van Houtte, E., Lebbe, L. (2008b). Study of the feasibility of an aquifer storage and recovery system in a deep aquifer in Belgium. *Hydrological Sciences Journal* 53(4), 844–856.

Wagner, V. (2010). Analysis of thermal response tests using advanced analytical and high resolution numerical simulations, Diploma thesis, University of Tübingen, Tübingen, Germany, 88 pp.

Wagner, R., Clauser, C. (2005). Evaluating thermal response tests using parameter estimation for thermal conductivity and thermal capacity. *Journal of Geophysics and Engineering* 2(4), 349–356.

Wagner, R., Rohner, E. (2008). Improvements of thermal response tests for geothermal heat pumps. *IEA Heat Pump Conference*, Zürich, Switzerland.

Wagner, V., Bayer, P., Kübert, M., Blum, P. (2012). Numerical sensitivity study of thermal response tests. *Renewable Energy* 41, 245–253.

Wagner, V., Blum, P., Kübert, M., Bayer, P. (2013). Analytical approach for groundwater-influenced thermal response tests of grouted borehole heat exchangers. *Geothermics* 46, 22–31.

Wagner, V., Li, T., Bayer, P., Leven, C., Dietrich, P., Blum, P. (2014). Thermal tracer testing in a heterogeneous sedimentary aquifer: Field experiment and numerical simulation. *Hydrogeology Journal.* (accepted)

Witte, H.J.L. (2001). Geothermal response test with heat extraction and heat injection: Examples of application in research and design of geothermal ground heat exchangers. *European Workshop on Geothermal Response Tests*, Lausanne, Switzerland.

Witte, H.J.L. (2013). Error analysis of thermal response tests, *Applied Energy* 109, 302–311.

Witte, H.J.L., van Gelder, G.J. (2006). Geothermal response tests using controlled multipower level heating and cooling pulses (MPL-HCP): quantifying ground water effects on heat transport around a borehole heat exchanger. *Proceedings of the Tenth International Conference on Thermal Energy Storage*, New Jersey, USA.

Witte, H.J.L., van Gelder, G.J., Spitler, J.D. (2002). In-situ measurement of ground thermal conductivity: The Dutch Perspective. *ASHRAE Transactions Research* 108(1), 263–273.

Wu, X., Pope, G.A., Shook, G.M., Srinivasan, S. (2008). Prediction of enthalpy production from fractured geothermal reservoirs using partitioning tracers. *International Journal of Heat and Mass Transfer* 51(5–6), 1453–1466.

Xue, Y., Xie, C., Li, Q. (1990). Aquifer thermal energy storage: A numerical simulation of field experiments in China. *Water Resources Research* 26(10), 2365–2375.

Yavuzturk, C., Spitler, J.D. (1999). Short time step response factor model for vertical ground loop heat exchangers. Paper presented at *ASHRAE Transactions* 105(2), 475–485.

Chapter 7

Case studies

In Chapter 6, two different field methods such as thermal response tests (TRTs) and short-term thermal tracer tests (TTTs) for the investigation of closed and open geothermal systems were discussed. Here the focus is on **long-term injection-storage and recovery experiments** and **regional studies on thermal use**.

The experiment on **aquifer storage and recovery of heated water** by Molz et al. (1978) at the **Auburn site near Mobile** (Alabama, USA) was numerically simulated and compared with field data by Papadopulos and Larson (1978). The results confirmed the utility of the simulation tools. Based on their analysis, the experimental techniques were improved by Molz et al. (1981) and successfully modeled by Tsang et al. (1981). The experiment consisted of the injection of 55,000 m³ of water of a temperature of 55.2°C into the aquifer of ambient temperature of 20°C. An injection period of 79 days was followed by a recovery phase of 52 days. The simulated production temperature of 32.8°C and the recovery rate (66%) agreed well with the observations. Parr et al. (1983) performed a field determination of the thermal energy storage parameters for the Mobile aquifer. Geochemical, thermal, and hydrogeological parameters were estimated by laboratory and field studies for the aquifer and the confining layers. The investigated parameters were the regional flow gradient, the vertical and horizontal hydraulic conductivities of the aquifer, the horizontal dispersivity, the vertical hydraulic conductivity of the confining layers, the thermal conductivities, heat capacities, and chemical characteristics of the aquifer matrix, and the groundwater.

Sauty et al. (1982b) performed field experiments of **warm water storage in a sandy gravel aquifer** confined by clay layers at the **Bonnaud** (Jura) site (France). The injected water volumes ranged from 500 to 1700 m³. The injection temperature ranged between 32.5°C and 40°C. Temperature profiles were observed in 17 boreholes. The results were discussed and were used to calibrate two mathematical models. A two-dimensional axisymmetric finite difference code (Sauty et al. 1982a) was used to determine the mean values of the parameters thermal conductivity, heat capacity, and thermal

dispersivity of four layers (Table 6.2). A three-dimensional finite difference model was used to determine their spatial variations in the horizontal
plane. In both models, density effects were disregarded. Finally, the results
of the storage experiments were successfully compared with the general type
curves for the production temperature, which were developed by Sauty et
al. (1982a).

Kobus and Söll (1992) modeled regional heat transport for two case studies in the shallow, unconfined, sandy gravel **Emme aquifer** close to **Kirchberg**
(Switzerland). The first case study concerned the injection of cold water at
the **Aefligen** (Switzerland) site (Blau et al. 1991). The test consisted of **local**
infiltration of 2 m^3 min^{-1}, at a temperature between 2°C and 7°C, in an
intermittent manner over a period of 150 days. The average infiltration rate
was 2 m^3 min^{-1}. The thermal plume was observed in several boreholes. It
was simulated by Söll (1985) with the support of a transient vertical model
along streamlines. With detailed consideration of the vertical velocity distribution and neglecting thermal dispersion, the comparison with field data
was satisfactory. However, when using depth-averaged velocities, thermal
macrodispersion effects, depending on the flow distance, were important.
The second case study was the **natural river water infiltration** from the river
Emme into the aquifer near Kirchberg (Blau et al. 1991). For the horizontal
modeling, two aquifer layers were necessary since a single depth-averaged
aquifer model was not able to account for the different flow directions in
both layers. The soil layer and the underlying impermeable layer were taken
into account by analytical solutions for the heat flux.

Thermal energy storage in an unconfined sand aquifer at the **Borden**
site (Canada) was investigated by Palmer et al. (1992). Heated water was
injected into a shallow aquifer, and thermal plume temperature was monitored in a dense array of piezometers. The total water volume of 53.5 m^3 of
temperature 35°C (with fluctuations between 34°C and 39°C) was injected
within 6 days. The initial temperature ranged between 9.5°C (at depth
6.1 m) and 15°C (depth 0.5 m). The detailed monitoring provided the three-
dimensional temperature distribution within the aquifer. The data were
used by Molson et al. (1992) to develop and validate a three-dimensional
coupled, density-dependent numerical model. The simulations provided
an excellent match between observations and the model. Thermal input
parameters, such as thermal dispersivity values, were obtained from either
literature data or from data analysis (Table 6.2).

Xue et al. (1990) performed simulations for three **seasonal aquifer ther-**
mal energy storage experiments conducted in **Shanghai** (China). They used
a three-dimensional flow and heat transport model. The aquifer used for
the heat storage was a confined sand formation. The results (temperature
and recovery rates) agreed well with the observations. They showed that
heat dispersion was important.

Birkholzer and Zhang (2000) modeled **hydrologic and thermal processes in a large-scale underground heater test in partially saturated fractured tuff at the Yucca Mountain site** (USA). They used the code TOUGH2 to simulate coupled water, water vapor, air, and heat transport in unsaturated fractured porous media. The agreement of the model results with long-term measurements indicated that the understanding of the complex process was satisfactory.

Markle et al. (2006) developed a **method for constructing the two-dimensional vertical thermal conductivity field** for a 12 m (horizontal) × 8 m (vertical) section of a glaciofluvial sand and gravel outwash deposit (**Tricks Creek study area**) in southwestern Ontario, Canada. The method used both field and laboratory measurements to determine the bulk thermal conductivity of the aquifer solids, the volumetric water content, and the porosity of the aquifer. Based on a model selection procedure using the information-theory approach, the Campbell model (Chapter 2.1.2.2) was selected as approximating the apparent thermal conductivity of variably saturated sands and gravels best. Thermal conductivities of the solid material of the aquifer were determined using two laboratory methods, first using the so-called divided-bar apparatus (Sass et al. 1971) and second using the mineral composition of the aquifer material. The mean thermal conductivity of fine to coarse sand was 4.22 ± 0.10 W m^{-1} K^{-1}, for gravel and sand it was 3.94 ± 0.12 W m^{-1} K^{-1}, and for till 3.72 ± 0.59 W m^{-1} K^{-1}. The two-dimensional vertical apparent thermal conductivity field was obtained by combining measured thermal conductivities and site stratigraphy with the measured porosity values. The resulting thermal conductivity field was used as input to a transient numerical model for simulating heat transport. In the saturated zone, the mean value and standard deviation of apparent thermal conductivity were 2.42 and 0.13 W m^{-1} K^{-1}, respectively. The apparent thermal conductivities in the unsaturated zone were between 40% and 50% lower than the apparent thermal conductivities in the saturated zone. Porosity strongly influenced the predicted two-dimensional conductivity field, indicating that this parameter has to be defined carefully. Numerical simulations were performed using the finite element numerical model HEATFLOW (Molson et al. 1992) after modifications were made to include the Campbell model for apparent thermal conductivity. Density effects on flow were taken into account. Numerical simulations were performed for a transport time of 10 days for both heterogeneous and average thermal conductivity fields. In both cases, uniform hydraulic conductivity was assumed. The simulations showed that using a homogeneous thermal conductivity instead of a fully heterogeneous field would yield temperature differences of less than 1 K relative to the heterogeneous cases. The authors concluded that, whenever small temperature differences are important, consideration of the heterogeneities in thermal conductivity may be necessary.

Lo Russo and Civita (2009) performed numerical investigations using FEFLOW for open-loop heat pumps of a **field study near Torino** (Italy). They tested scenarios with respect to the environmental impact.

Nam and Ooka (2010) performed numerical heat transfer simulations for their case study at the **Chiba experimental station** in Tokyo (Japan). The simulation model was confirmed by the experimental results. Based on the model, several methods for achieving an optimal coefficient of performance were tested.

Kupfersberger (2009) investigated the **impact of groundwater heat pumps** in the shallow sandy gravel aquifer Leibnitzer Feld (Leibnitz close to Graz, Austria) with the help of a two-dimensional numerical flow and heat transport model using the FEFLOW software. Transient head data were used to calibrate the flow model. A heat-transfer function between soil surface (using air temperature) and groundwater was established to represent the heat input rate. The heat transfer coefficients (transfer-in and transfer-out parameters) were calibrated using observed temperature profiles. The temperature of the reinjected water was limited to a decrease of 5 K and an absolute minimum of 5°C by Austrian law. The limited temperature fluctuations justified neglecting density effects in the flow modeling. The annual heat extractions of three users were assumed to be 116, 321, and 600 MWh, respectively. It was shown that the reduction of the groundwater temperature 300 m downstream of the reinjection wells was less than 0.5°C in these cases, hence demonstrating the feasibility and potential of the thermal use.

Vandenbohede et al. (2011) studied the heat transport during a **shallow heat injection and storage field test at the Ghent University site** in Ghent (Belgium). The seasonal temperature fluctuations were simulated using the code SEAWAT. They found that the most sensitive parameter is the thermal conductivity of the solid phase followed by the porosity, the heat capacity of the solid, and the longitudinal dispersivity. They demonstrated the dominance of conductive transport during the storage phase, while the convective transport was dominant during the injection phase. They concluded that dispersivity cannot be ignored in the simulation of advective heat transport in aquifers.

In the following, two regional case studies from Austria and Switzerland and one local case study of a ground source heat pump system in Germany are presented in more details.

7.1 CASE STUDY ALTACH (AUSTRIA)

The **change of the thermal regime due to the thermal use of the aquifer in the Rhine valley in the region of Altach in the State of Vorarlberg** (Austria) was investigated using a two-dimensional numerical model (Cathomen

2002; Cathomen et al. 2002). The task consisted of evaluating the impact of existing installations of open systems with and without heat pumps and the warming effects of buildings (warm basements) on the groundwater temperature. The motivation was to compare long-term effects with local temperature measurements in order to assess a possible overexploitation of the thermal resource. In the State of Vorarlberg, the thermal use of aquifers using heat pumps has been promoted and supported since the 1980s. For this purpose, the thermal potential of the regional aquifer was assessed by the State authorities, with the help of a water and heat balance over sub-regions. Based on these investigations, licenses for thermal use have been issued by the authorities. In those times, the State of Vorarlberg limited the allowed maximum change of the temperature in aquifers to 1 K. Since measurements indicated that the maximum temperature decrease reached about 1 K locally, the question of whether the maximum potential is already exhausted arose. However, meanwhile, the authorities of Austria fixed the maximum allowable change to 6 K, relative to the existing groundwater temperature at the location of the reinjection (ÖWAV 2009; Chapter 1.8.2). Thus, the situation is uncritical now.

The **town and the region of Altach** are located about 20 km south of Bregenz, the capital of the State of Vorarlberg, within the **plain of the Rhine valley** at an altitude of 412 m a.s.l. The **boundaries of the investigation area** were chosen such that no thermal plumes from neighboring regions were included. The resulting domain was about 21 km^2 in size. It was delimited at the western boundary by River Rhine (flowing from southwest to northeast) and Alter Rhein, which is an oxbow lake of the old course of the river. The northeastern and the southern boundaries were chosen along steady-state streamlines. The northeastern boundary as well as part of the southwestern boundary delimit the highly permeable region, which borders hilly and mountainous regions. The domain contains eight groundwater pumping stations and several creeks. The **annual mean air temperature** (2 m above soil surface) is 10.1°C, and the **mean temperature of the soil surface** (in 5 cm depth) is about 11.2°C. These data were measured in 1999 at the meteorological station of Vaduz (Principality of Liechtenstein), which is located about 30 km south of Altach up-valley at an altitude of 460 m a.s.l., and which is run by MeteoSwiss. In 1999, precipitation was 1297 mm. The average temperature of River Rhine (Hydrological Annual Book of Switzerland, 2000, station Diepoldsau) was 8.1°C in the year 2000 and therefore cooler than air and soil surface temperature (7.8°C in the period 1984–2000). However, since groundwater is essentially outflowing into River Rhine, only limited cooling effects are expected.

Geologically, the Rhine valley is essentially filled with unconsolidated rock material, which was deposited into an elongated lake during the last ice age. Fine-grained sediments form the bottom of a sandy gravel aquifer, which represents the highly conductive top layer. The aquifer thickness is typically 15–20 m, and the depth to groundwater is 2–4 m. Hydraulic conductivity values are on the order of 2×10^{-3} m s^{-1}. Furthermore, two large alluvial cones were deposited by creeks from hills and mountains. In these subregions, the altitude and thus the depth to groundwater are increased accordingly.

The domain contains the **communities and towns of Altach, Mäder, and Hohenems** with a total population of about 30,000 inhabitants. Land use in the region is mainly agriculture (about 50% of the area) and settlements as well as industry. Data of the State of Vorarlberg allowed a quantification of the thermal input and extractions (Table 7.1). A total of 323 (2002) **installations with groundwater heat pump systems** with a total annual mean performance of 1.35 MW was registered. Concerning the groundwater temperature, the Hydrographic Service of the State of Vorarlberg performed **periodic or permanent temperature measurements** in several boreholes. These measurements were taken routinely at a fixed depth of 1 m below the lowest expected groundwater table. The annual mean groundwater temperature of five stations was 11.5°C with mean values at the individual stations between 10.4°C and 11.9°C. These data were complemented by the **measurement of temperature profiles** in a total of 17 piezometers within the domain, with the help of the Hydrographic Service (Figure 7.1). The temperature profiles were measured in late fall (November 13, 2001) using a thermal borehole probe, in intervals of 1 m, starting at the water table. Most of the temperature profiles follow the characteristic shape of a damped harmonic wave according to Equation 3.85 for one-dimensional vertical conductive heat transport for this date. The maximum temperature was observed at a depth of about 4–5 m below soil surface. Most of the profiles tend toward the annual mean temperature of approximately 11°C at a depth of about 10 m. Exceptions are stations 50.4.07 and Kropf-2. The former profile was possibly influenced by vertical flow in the borehole, with inflowing groundwater at larger depths into and outflowing groundwater at higher depths out of the borehole. The piezometer of the latter profile is situated close to a groundwater lake (formed by gravel exploitation) with stagnant water, which can possibly influence the groundwater temperature. The variability of the rest of the temperature profiles can be attributed to local land use and the thermal properties of the local material, but also by the thermal use of the aquifer.

Table 7.1 Altach study: annual total heat flux

Thermal input category	Thermal input [GWh]
Re-injection of warm water from heat exchange	3.2
Re-injection of cool water from heat pumps	−11.9
Thermal inflow from warm basements	6.3
Total thermal input	−2.4

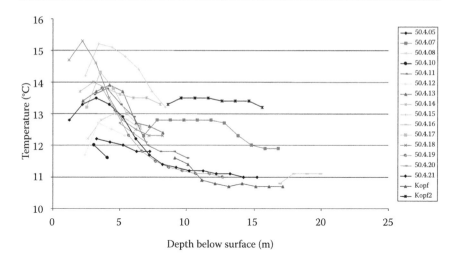

Figure 7.1 **(See color insert.)** Altach study: temperature profiles in boreholes, measured on November 13, 2001. (Modified after Cathomen, N., Wärmetransport im Grundwasser, Auswirkungen von Wärmepumpen auf die Grundwassertemperatur am Beispiel der Gemeinde Altach im Vorarlberger Rheintal. Diploma thesis, ETH Zurich, Institute of Hydromechanics and Water Resources Management, 2002.)

For the **simulation of the regional heat transport**, it was conceived that only **temperature change** ΔT, which is caused by anthropogenic influence, relative to the "natural" conditions, is relevant and thus modeled. Therefore, **heat input** is considered by the following sources:

- **Reinjection of warm water from heat exchange** for cooling purposes
- **Reinjection of cool water from heat pumps** for heating purposes
- **Thermal inflow from warm basements** of buildings

A further heat input is due to the sewer system and possibly other infrastructure devices, but those were disregarded here. The energy extraction by heat pumps is relatively well known from the licensing documents. For the injection of warm water, only the discharge was known. The related heat load was estimated as a mean temperature increase of 4 K. The heat inflow from basements was estimated using a typical basement temperature of 15°C and assuming a linear mean temperature gradient between basement and average groundwater depth, together with a typical thermal conductivity value. The area of the basements was determined from the fraction of the buildings in the settlement area. This fraction was estimated to be 10% based on the topographic map. The balance of the anthropogenic heat input is shown in Table 7.1. From the balance, it can be seen that the estimated total heat input is negative, indicating a net cooling effect over the area.

For the simulation, the **software package** PMWIN (Chiang and Kinzelbach 2001) was used. The flow calculations were performed with the module MODFLOW96 (USGS 2012) and the **transport calculations** with module MT3DMS (Zheng and Wang 1999). The model was conceived as a two-dimensional, unconfined, one-layer model. Therefore, the variables head and temperature correspond to average values over the thickness of the aquifer. The flow model was conceived for steady-state flow conditions, thus neglecting seasonal variations of boundary conditions like inflow rates. The resulting flow velocities are therefore long-term average values. With regard to the relatively small, expected heat transport velocities of the order of 200 m year^{-1}, and the small depth to groundwater (2–3 m) and the relatively small aquifer thickness (15–20 m) in the relevant part of the model, this procedure is justified. The **hydraulic conductivity field and the lateral inflow rates** were taken from the existing and calibrated steady-state groundwater model of the International Governmental Commission of the Alpine Rhine region (IRKA). The size of the finite difference cells was 100 m. The recharge rate due to precipitation was 0.5 mm day^{-1}. This value was estimated considering the time span before the measurement campaign. Fixed head boundary conditions were set along River Rhine and Alter Rhein based on measurements. Along the northeastern boundary as well as part of the southwestern boundary, lateral inflow was prescribed based on flux estimates. Interactions with **creeks** were modeled by the leakage concept. **Head isolines** are presented in Figure 7.2. The flow in the plain area is mainly parallel to River Rhine with a typical flow gradient of about 0.002. Comparison of the calculated heads with measured data (IRKA) yielded a standard deviation of 0.5 m.

The **heat transport simulation** was performed according to the mathematical formulation presented in Chapter 4.1 for **two-dimensional aquifers**. By **utilizing the analogy between solute transport and heat transport** (Chapter 4.1.1), the transport parameters of the two-dimensional MT3DMS model were determined. As outlined above, the temperature difference $\Delta T(\mathbf{x})$ was modeled. This implies that also heat transport is considered as steady state. This was achieved using the transient transport code MT3DMS over a sufficiently long time span. The initial temperature increment $\Delta T(\mathbf{x}, t = 0)$ was set to zero. For streamline boundaries, a thermally insulating boundary was assumed. Inflow boundaries were set to zero thermal inflow boundaries. The influence of deeper sediment layers and a geothermal heat flux were neglected in the thermal model. Longitudinal dispersivity was chosen as 50 m and the transversal one as 5 m. Porosity was estimated as 0.25, λ_m as 2.7 W m^{-1} K^{-1}, C_m as 2.87×10^{-6} J m^{-3} K^{-1}, and the effective vertical thermal conductivity $\lambda_{eff,vert}$ as 1.8 W m^{-1} K^{-1}. The equivalent transport coefficients were calculated cell-wise using areal information on the land

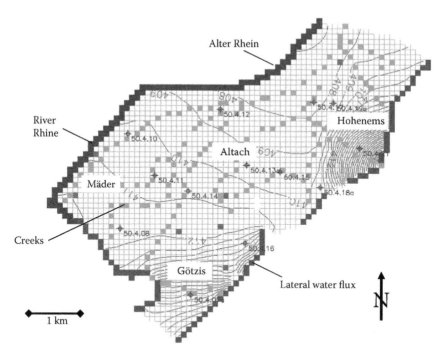

Figure 7.2 **(See color insert.)** Altach study: two-dimensional flow model Altach (Austria) with head isolines (equidistance 1 m). Dark blue: river cells; blue: prescribed inflow cells; bright blue: creeks. (Modified after Cathomen, N., Wärmetransport im Grundwasser, Auswirkungen von Wärmepumpen auf die Grundwassertemperatur am Beispiel der Gemeinde Altach im Vorarlberger Rheintal. Diploma thesis, ETH Zurich, Institute of Hydromechanics and Water Resources Management, 2002.)

use (i.e., basement) depth to groundwater, aquifer thickness, and thermal parameters of the subsurface. The areal thermal inflow from warm basements of buildings was estimated cell-wise as described above. Initial relative temperature as well as boundary conditions (lateral inflow) were set to zero everywhere.

Simulations were performed for 10 and 75 years. It was shown that the thermal plumes can be considered in quasi steady state after about 10 years. The temperature differences were calculated separately for the different thermal input cases (reinjection of warm water from heat exchange, reinjection of cold water from heat pumps, warming by basements). The reason for this procedure is that in the analogy to solute transport, the concentrations have to be positive. Therefore, all cases were calculated with positive absolute values of temperature differences separately, and the results were finally superimposed. The **resulting distribution of the relative temperature** $\Delta T(\mathbf{x})$ caused by reinjection of warm water from heat exchange, reinjection from groundwater heat pumps, and thermal inflow from warm basements is shown in Figure 7.3.

Simulated maximum and minimum temperature changes were 3.2 and –2.6 K, respectively. Peak regions are found close to and downstream of settlements mainly. Note that warming plumes occur as often as cold plumes, and both are of similar absolute value. For the inner settlement area of the town of Altach, it was found that the temperature change was close to zero or even slightly positive. This means that the thermal input in this subregion is almost balanced. The **simulated temperature differences are compared with the measured** ones (Table 7.2). Absolute differences between both are smaller than 1 K. We have to keep in mind that the measured data

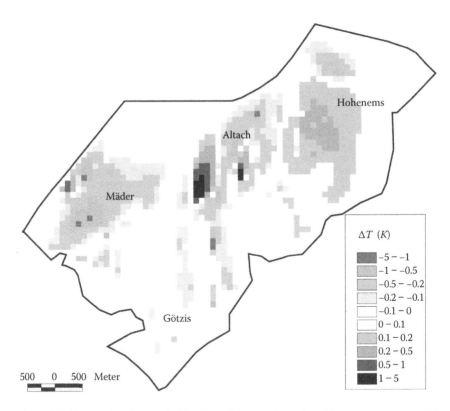

Figure 7.3 **(See color insert.)** Altach study: two-dimensional heat transport model Altach (Austria) with temperature increase due to thermal use (groundwater heat pumps, heating by constructions). Dark blue: river cells; blue: prescribed inflow cells. (Modified after Cathomen, N., Wärmetransport im Grundwasser, Auswirkungen von Wärmepumpen auf die Grundwassertemperatur am Beispiel der Gemeinde Altach im Vorarlberger Rheintal. Diploma thesis, ETH Zurich, Institute of Hydromechanics and Water Resources Management, 2002.)

Table 7.2 Altach study: measured and simulated mean temperature differences $\Delta T = T - T_{mean}$

Station	Depth of measurement below soil surface (m)	Measured mean temperature difference (K)	Simulated mean temperature difference (K)
50.4.11	4.85	+0.70	−0.2
50.4.13	4.33	−0.26	+0.0
50.4.21	3.20	+0.81	+0.1
50.4.17	3.92	−0.85	+0.2
50.4.20	5.90	+0.24	−0.1
Average		+0.47	+0.0

are measured 1 m below the deepest groundwater table level. Therefore, the measurements do not necessarily represent average profile temperatures exactly. Nevertheless, it can be concluded that the model simulates the mean temperatures with an accuracy of about 0.5 K. The inaccuracy includes the effect of variable soil surface temperature over the area.

7.2 LIMMAT VALLEY AQUIFER ZURICH (SWITZERLAND)

The **Limmat Valley aquifer** extends from the downtown area of the city of **Zurich** (Switzerland) and River Sihl along River Limmat over a distance of about 16 km with a mean and maximum width of about 1 to 2 km. The total area is about 20 km² comprising the part with aquifer thickness larger than 2 m. It is a highly conductive, unconfined, sandy gravel aquifer. The Limmat Valley was formed in its main disposition in the early Pleistocene (Kempf et al. 1986). The main erosion of the U-shaped valley occurred during the Riss ice age. The sediment filling occurred mainly in the Würm ice age with several stages of the glaciation and related forming of moraines and fluvial deposits. Therefore, the structure of the aquifer is quite complex and consists of sand, gravel deposits, lake sediments, and moraine material. The contributing rivers are River Limmat, starting downtown Zurich at the lake of Zurich, and the smaller River Sihl, which extends from the perialpine region to its confluence with River Limmat. Land surface topography ranges between 380 and 410 m a.s.l.

River Sihl is infiltrating into the aquifer via unsaturated percolation with a mean annual temperature of 10.7°C–11.3°C (period 2000–2003). However, the expected infiltration rates are relatively small. On the other hand, infiltration from **River Limmat** is more significant. It occurs along the major part of the river section mostly directly into the aquifer, in parts via unsaturated percolation. The corresponding **annual temperature** is 10.2°C–11.3°C (period 2000–2003). River water infiltration influences the seasonal temperature regime considerably, with its amplitude decreasing with distance from the river. In the downtown area, mean groundwater temperatures range between 12.1°C and 16.5°C (Jäckli 2010). **How can the increase in the groundwater temperature due to anthropogenic influence roughly be assessed?** Based on the annual air temperature of the station Zurich-Affoltern of 9.6°C–10.3°C (period 2000–2003, data MeteoSwiss) and the soil surface temperature of 11.3°C–11.9°C with a mean value of 11.5°C, the mean difference between the air and soil temperature amounts to 1.2–1.7 K with a mean value of 1.5 K. Adopting roughly the temperature difference between soil surface and noninfluenced groundwater (outside the range of infiltrating rivers), the temperature increase due to anthropogenic reasons is therefore between about 0.5 and 5 K, which is the same range as reported in German cities (Menberg et al. 2013). This estimated anthropogenic increase depends mainly on the density of warm basements and the depth to groundwater. Outside the downtown region of Zurich, the anthropogenic influence is expected to be smaller. Nevertheless, a typical temperature increase of 2–3 K in urban areas is also postulated by the groundwater agency of the Canton of Zurich (AWEL 2005). This represents a considerable additional potential for thermal use, besides the allowed temperature change of 3 K according to the Swiss regulations. Systematic, long-term groundwater temperature monitoring does not yet exist for the Limmat Valley aquifer but is planned for the future (AWEL 2005). Note that the temperature of River Limmat is slightly higher than the estimated soil surface temperature and that River Sihl is clearly cooler.

The authorities of the Canton of Zurich were relatively reluctant in the past with respect to the promotion of the thermal use of groundwater. Nevertheless, about 100 installations have been approved. As a rule, the **performance of an installation** has to exceed 150 kW in order to ensure improved quality standards in maintenance. For the future, an increase in the thermal use of groundwater is expected.

In the context of master projects at the Institute of Environmental Engineering, ETH Zurich, **groundwater temperature profiles** of 47 piezometers in the Limmat Valley aquifer were measured on April 14, 2005. They show the vertical temperature profiles typical for the season and the influence of the infiltration of river water. Maximum temperatures were mainly found at greater depth. These values, lying outside of the influence of river water infiltration, can be considered as indicators of the mean groundwater temperature. In the downtown area, these values range between 12.2°C and 14.6°C, thus similar to the data of Jäckli (2010).

For various sections of the Limmat Valley aquifer, **two-dimensional groundwater flow and heat transport modeling** was performed. The method and codes correspond to those already described for the Altach study

(Section 7.1). For illustration, results are shown for the **region of the town of Schlieren** (Müller and Ott 2005; Figure 7.4). This town is situated down-valley from the City of Zurich. The area of the model domain was 4.7 km². The area comprises settlement areas (including industry) with a fraction of 60%. The rest consists of agricultural areas as well as recreational areas. The domain contains nine groundwater pumping stations for drinking and process water (total of about 6300 m³ day⁻¹). Boundary condition types of the flow model are shown in Figure 7.4. Measured hydraulic conductivity values (2×10^{-4} to 10^{-2} m s⁻¹) were interpolated, and the recharge rate was estimated. Leakage coefficients of River Limmat and lateral inflow rates were calibrated. Comparison between calculated and measured heads (on April 14, 2005) yielded a mean standard deviation of about 0.3 m. For simplicity, it was assumed that the mean temperature of the river water corresponds to the "natural" groundwater temperature. Heat flow from basements to groundwater was calculated using Equation 4.5 with a temperature difference of 3 K with respect to the soil surface temperature and the areal fraction of buildings in the finite difference cells. The **simulated warming effect by basements alone** is shown in Figure 7.5. The results seem plausible. Direct comparison with measured data was not possible as no data were available. The **superimposed effect of two possible installations for thermal use** of a capacity of 315 and 196 kW is shown in Figure 7.6. The results indicate a

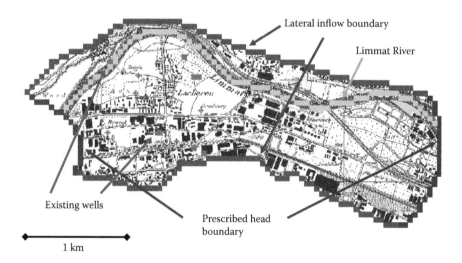

Figure 7.4 **(See color insert.)** Case study Limmat Valley, subregion town of Schlieren: model domain with wells, and boundary conditions. (Modified after Müller, E., Ott, D., Thermische Nutzung des Grundwasserleiters Limmattal, Teilgebiet Hardhof-Schlieren [Thermal use of the Limmat Valley aquifer: Area Hardhof-Dietikon]. Report Master Project, ETH Zurich, Institute of Environmental Engineering, 2005.)

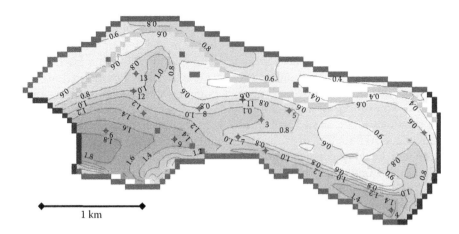

1 km

Figure 7.5 **(See color insert.)** Case study Limmat Valley, subregion town of Schlieren: quasi-steady-state simulation of the temperature increase by warm basements. (Modified after Müller, E., Ott, D., Thermische Nutzung des Grundwasserleiters Limmattal, Teilgebiet Hardhof-Schlieren [Thermal use of the Limmat Valley aquifer: Area Hardhof-Dietikon]. Report Master Project, ETH Zurich, Institute of Environmental Engineering, 2005.)

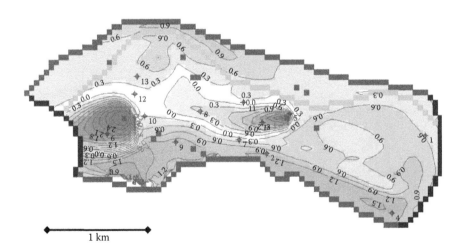

1 km

Figure 7.6 **(See color insert.)** Case study Limmat Valley, subregion town of Schlieren: quasi-steady-state simulation of the temperature increase by warm basements, and, superimposed, two planned installations for the thermal use of groundwater. (Modified after Müller, E., Ott, D., Thermische Nutzung des Grundwasserleiters Limmattal, Teilgebiet Hardhof-Schlieren [Thermal use of the Limmat Valley aquifer: Area Hardhof-Dietikon]. Report Master Project, ETH Zurich, Institute of Environmental Engineering, 2005.)

maximum temperature decrease of 4.5 K. The requirement of a maximum change in temperature downstream at a distance of 100 m according to the Swiss regulations (Chapter 1.8.1) is met for both installations. Moreover, an interference of the two installations with respect to groundwater temperature is avoided.

7.3 BAD WURZACH (GERMANY)

Only a few geothermal test sites for GSHP systems exist worldwide, which allow studying the thermal development in the subsurface in the vicinity of a borehole heat exchanger (BHE). A prominent example is the Elgg site, Switzerland, which is already mentioned in Chapter 5.1. Eugster (2001) and Rybach and Eugster (2010) monitored and simulated the long-term behavior of a BHE in a conduction-dominated system with negligible groundwater influence. The data showed that over the last years of the operation (1996–1998), the temperature stabilized showing only minor temperature changes (<1 K) in comparison to the beginning of the operation in 1986.

In the framework of the research project "geomatrix.bw," a **test site for a geothermal ground source heating pump** (GSHP) **system** was developed in **Bad Wurzach**, South Germany (Wagner and Blum 2012; Bisch et al. 2012). The main objective of the project was to monitor the temperature distribution of the near field of a BHE. However, in contrast to the study in Elgg, the examined BHE is partially **influenced by advection**. In 2009, a BHE (EW1/09) with double U-tubes and five observation wells (B0 at larger distance, B1–4 in the vicinity) were installed (Figure 7.7). Also, four coaxial systems were installed, but here we focus on the BHE with the four adjacent observation wells. The site is located south of a marshland in the Molasse basin. In the study area, the stratigraphy is characterized by Quaternary gravels, sands, and clays on top followed by clay-, marl-, and sandstones of the Upper Freshwater Molasse (UFM), which were encountered on site at 58 m depth, with a total thickness of 400–500 m. The upper part of the geological profile of the BHE (EW1/09) with a total length of 100 m based on core drilling is illustrated in Figure 7.8. The distances of the four downstream observation wells (B1, B2, B3, and B4) from the BHE are 3.9, 5.7, 7.9, and 14.1 m, respectively (Figure 7.7). The temperature sensors (Pt100) in the BHE and the observation wells were merely installed in the upper Quaternary aquifer and aquitard down to 35 m depth (Figure 7.8).

The local groundwater flow direction in the upper aquifer is generally north toward the marshland, which begins at a distance of about 400 m from the study site. Repeated and continuous groundwater level measurements showed that the groundwater flow direction fluctuates between 2° and 7° east and the groundwater level changes by a maximum of about 25 cm. The average hydraulic gradient is 0.002. In a nearby pumping well, a

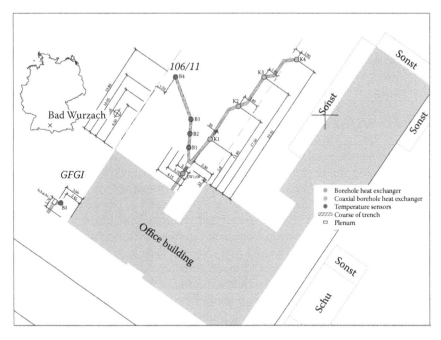

Figure 7.7 **(See color insert.)** Bad Wurzach study: site map showing the locations of the BHE (EW1/09) and the five observation wells (B0, B1, B2, B3, and B4).

long-term pumping test resulted in a hydraulic conductivity of 7×10^{-3} m s^{-1} (Table 7.3), which can be considered an average value for the heterogeneous aquifer.

Temperature monitoring in the observation wells over a time period of one year is illustrated in Figure 7.9. In winter 2011, the heating results in a minor decrease in subsurface temperatures (–1.1 K), which is more pronounced in the aquitard. However, this **thermally affected zone** (TAZ) is only visible in a distance of 3.9 m at the observation well B1. Furthermore, in summer 2011, the subsurface temperature indicates a full thermal recovery. The maximum temperature change due to the cooling and heating operation of the GSHP system was 2.9 K (7.7°C in March 2011 and 10.6°C in September 2012) at observation well B1 in 21 m depth (Bisch et al. 2013). At the observation well B4 (distance of 14.1 m), however, the maximum detected temperature change is only 0.4 K showing the limited spatial reach of the thermal disturbance around the BHE, despite the influence by a groundwater flow velocity of about 3.5 m day^{-1}.

To further study the evolution of the TAZ of the GSHP system, a 3D numerical heat transport model was set up in FEFLOW (Figure 7.10). For the latter, the input values, which are provided in Table 7.3, were used. Due to the incomplete data, a simplified layered model was implemented.

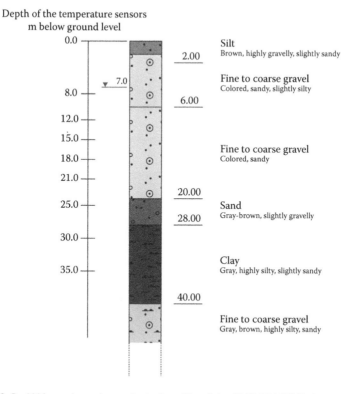

Depth of the temperature sensors
m below ground level

Depth	Layer depth	Description
	2.00	**Silt** Brown, highly gravelly, slightly sandy
	6.00	**Fine to coarse gravel** Colored, sandy, slightly silty
	20.00	**Fine to coarse gravel** Colored, sandy
	28.00	**Sand** Gray-brown, slightly gravelly
	40.00	**Clay** Gray, highly silty, slightly sandy
		Fine to coarse gravel Gray, brown, highly silty, sandy

Figure 7.8 Bad Wurzach study: geological profile of the BHE (EW1/09) determined by drilling cuttings and depth of the installed temperature sensors in the observation wells B1, B2, and B3.

Table 7.3 Determined, estimated, and fitted hydraulic and thermal parameters of the geothermal test site at Bad Wurzach, Germany

Parameter	Value
Hydraulic conductivity, K_w (m s^{-1})	0.007[a]
Hydraulic gradient (hor.), I_{hor} (–)	0.002[b]
Porosity, ϕ (–)	0.35[c]
Longitudinal dispersivity, β_L (m)	5.0[d]
Transverse dispersivity, β_T (m)	5.0[d]
Volumetric heat capacity of the porous media, C_m [J (m^3 K)$^{-1}$][c]	2.4×10^6
Thermal conductivity of the porous media, λ_m [W (m K)$^{-1}$]	1.4[c]

[a] Long-term pumping test at the Haidgauer Haide from 1968 (Weinszieher 1984).
[b] Hydraulic gradient based on a determined groundwater contour map (Wagner and Blum 2012).
[c] Estimated parameter value.
[d] Fitted parameter value.

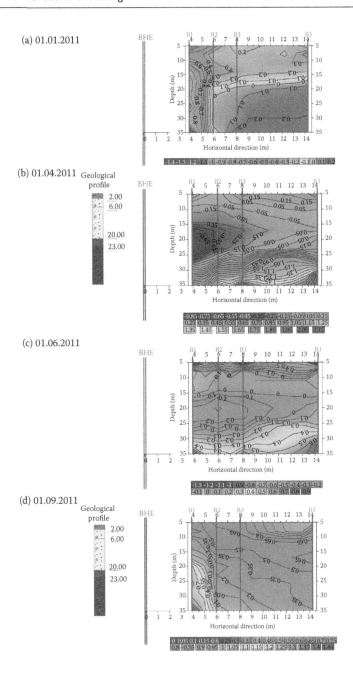

Figure 7.9 **(See color insert.)** Bad Wurzach study: cross section of the temperature measurements in various depths in groundwater flow direction over a time period of one year.

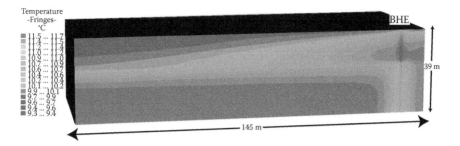

Temperature
-Fringes-
°C

Figure 7.10 **(See color insert.)** Bad Wurzach study: 3D numerical heat transport model of the BHE (EW1/09) showing the resulting temperature plume after 160.4 days.

The time period between May 13, 2011 and December 6, 2011 (200 days) was chosen for a preliminary modeling study, because before that period, no heat meter was installed at the heat pump and therefore no detailed information on heat extraction by the BHE was available. The measured temperatures at the observation wells for the studied time period at a depth of 21 m are shown in Figure 7.11. The maximum observed temperature difference in this time period is smaller than 2.5 K in the closest observation

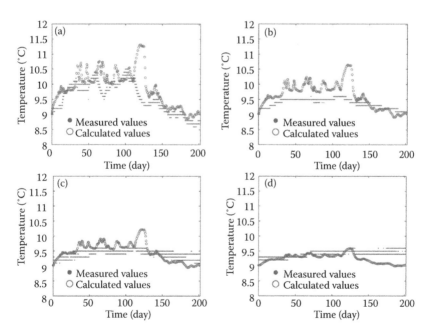

Figure 7.11 **(See color insert.)** Bad Wurzach study: comparison between measured and simulated temperatures at the four observation wells (a) B1, (b) B2, (c) B3, and (d) B4 in 21 m depth in the vicinity of the BHE (EW1/09).

well B1. The recorded temperature developments in all wells clearly show the cooling period in spring and summer, and the heating period in autumn and winter, reflected by increase and decrease in groundwater temperatures. As expected, this temperature signal is strongly damped with increasing distance from the BHE. After manual calibration of the dispersivity, the results of the preliminary numerical simulations roughly match the monitored temperatures (Figure 7.11). However, this fit could only be obtained using a relatively large dispersivity value of 5 m. This shows that hydraulic heterogeneity causes substantial macrodispersive spreading of the thermal plume. The model that assumes a homogeneous aquifer layer can only approximate these effects by specifying a large dispersivity (Table 7.3). The down-gradient observation wells show smoother temperature trends than the simulated ones, which could potentially be captured by an increasing dispersivity. Alternatively, this could also indicate an underestimation of the role of heat diffusion or temporal variability of the flow field.

Both geothermal test sites, in Elgg (Switzerland) and in Bad Wurzach (Germany), with mainly conductive or partially advective conditions, indicate that the TAZs around BHEs are spatially limited even after longer operation time. Hence, any planned or required temperature monitoring, for example, in the context of licensing issues by environmental authorities (Hähnlein et al. 2013), should carefully consider these circumstances.

REFERENCES

AWEL (2005). *Massnahmenplan Wasser im Einzugsgebiet Limmat und Reppisch (Planned Measures for the Catchment of Rivers Limmat and Reppisch)*. Amt für Abfall, Wasser, Energie, Luft, Kanton Zürich, Zurich, Switzerland.

Birkholzer, J.T., Tsang, Y.W. (2000). Modeling the thermal-hydrologic processes in a large-scale underground heater test in partially saturated fractured tuff. *Water Resources Research* 36(6), 1431–1447.

Bisch, G., Klaas, N., Braun, J. (2012). geomatrix.bw: Teil 3: Validierung von Erdwärmesondensimulationen zum Kühlen und Heizen im Nah- und Fernfeld mit Hilfe geothermischer Testfelder. Report ZO4E28002, Stuttgart.

Blau, R.V., Werner, A., Würsten, M., Kobus, H., Söll, T., Zobrist, J., Hufschmied, P. (1991). *Natürlicher und künstlicher Wärmeeintrag: Auswirkungen auf den Grundwasserhaushalt (Natural and Experimental Temperature Effects: Consequences on Ground Water Resources)*. Gas, Wasser, Abwasser (GWA), Schweizerischer Verein des Gas- und Wasserfaches, Zürich, Switzerland, Vol. 71, pp. 173–230, ISSN 0036-8008.

Cathomen, N. (2002). Wärmetransport im Grundwasser, Auswirkungen von Wärmepumpen auf die Grundwassertemperatur am Beispiel der Gemeinde Altach im Vorarlberger Rheintal. Diploma thesis, ETH Zurich, Institute of Hydromechanics and Water Resources Management, Zurich, Switzerland.

Cathomen, N., Stauffer, F., Kinzelbach, W., Osterkorn, F. (2002). Thermische Grundwassernutzung. Auswirkung von Wärmepumpenanlagen auf die Grundwassertemperatur. *GWA* 82(12), 901–906.

Chiang, W.-X., Kinzelbach, W. (2001). *2D Groundwater Modeling with PMWIN: A Simulation system for Modeling Flow and Transport Processes*. Springer, Berlin.

Eugster, W.J. (2001). *Langzeitverhalten der EWS-Anlage in Elgg (ZH)—Spotmessung im Herbst 2001*. Swiss Federal Office of Energy, Bern, Switzerland, 14 pp.

Hähnlein, S., Bayer, P., Ferguson, G., Blum, P. (2013). Sustainability and policy for the thermal use of shallow geothermal energy. *Energy Policy* 59, 914–925.

Jäckli. (2010). *Wohnsidlung Hardau II, Bullingerstrasse, Grundwasser-Wärmenutzung, Numerische Modellierung von bestehenden und künftigen Nutzungen im Stadtgebiet Zürich (Numerical Modeling of Existing and Future Thermal Use in the City Area of Zurich)*. Dr. Heinrich Jäckli AG, Zürich, Switzerland; Stadt Zürich, Immobilien-Bewirtschaftung, Zürich, Switzerland (unpublished).

Kempf, T., Freimoser, M., Haldimann, P., Longo, V., Müller, E., Schindler, C., Styger, G., Wyssling, L. (1986). Die Grundwasservorkommen im Kanton Zürich. Beiträge zur Geologie der Schweiz, Geotechnische Serie. Issued by: Schweiz. Geotechnische Kommission and Direktion der öffentlichen Bauten des Kantons Zürich.

Kobus, H., Söll, T. (1992). Modellierung des regionalen Wärmetransports: Fallbeispiele Kaltwassereinleitung Aefligen und Emmeinfiltration Kirchberg (Modelling of regional heat transport: Case studies coldwater discharge Aefligen and Emme infiltration Kirchberg). In: H. Kobus (Ed.), *Schadstoffe im Grundwasser, Band 1: Wärme- und Schadstofftransport im Grundwasser*. Deutsche Forschungsgemeinschaft DFG, VDH Verlagsgesellschaft mbH, Weinheim, Germany, pp. 341–375.

Kupfersberger, H. (2009). Heat transfer modeling of the Leinitzer Feld aquifer (Austria). *Environmental Earth Science* 59, 561–571.

Lo Russo, S., Civita, M.V. (2009). Open-loop heat pumps development for large buildings: A case study. *Geothermics* 38, 335–345.

Markle, J.M., Schincarion, R.A., Sass, J.H., Molson, J.M. (2006). Characterizing the two-dimensional thermal conductivity distribution in a sand and gravel aquifer. *Soil Science Society of America Journal* 70, 1281–1294.

Menberg, K., Bayer, P., Zosseder, K., Rumohr, S., Blum, P. (2013). Subsurface urban heat islands in German cities. *Science of the Total Environment* 442, 123–133.

Molson, J.W., Frind, E.O., Palmer, C.D. (1992). Thermal energy storage in an unconfined aquifer. 2. Model development, validation, and application. *Water Resources Research* 28(19), 2857–2867.

Molz, F.J., Parr, A.D., Andersen, P.F. (1981). Thermal energy storage in a confined aquifer. *Water Resources Research* 17(3), 641–645.

Molz, F.J., Warman, J.C., Jones, T.E. (1978). Aquifer storage of heated water: Part I—A field experiment. *Ground Water* 16(4), 234–241.

Müller, E., Ott, D. (2005). Thermische Nutzung des Grundwasserleiters Limmattal, Teilgebiet Hardhof-Schlieren (Thermal useofthe Limmat Valleyaquifer: Area Hardhof-Dietikon). Report Master project, ETH Zurich, Institute of Environmental Engineering, Zurich, Switzerland.

Nam, Y., Ooka, R. (2010). Numerical simulation of ground heat and water transfer for groundwater heat pump system based on real-scale experiment. *Energy and Buildings* 42(1), 69–75. http://journals.ohiolink.edu/ejc/article.cgi?issn=0 3787788&issue=v42i0001&article=69_nsoghapsbore.

ÖWAV (2009). Thermische Nutzung des Grundwassers und des Untergrunds— Heizen und Kühlen (Thermal use of groundwater and underground— Heating and cooling). ÖWAV-Regelblatt 207, Österreichischer Wasser- und Abfallwirtschaftsverbands ÖWAV (Guideline 207 of the Austrian Water and Waste Management Association), Vienna, Austria.

Palmer, C.D., Blowes, D.W., Frind, E.O., Molson, J.W. (1992). Thermal energy storage in an unconfined aquifer. *Water Resources Research* 28, 2845–2856.

Papadopulos, S.S., Larson, S.P. (1978). Aquifer storage of heated water: Part II— Numerical simulation of field results. *Ground Water* 16(4), 242–248.

Parr, A.D., Molz, F.J., Melville, J.G. (1983). Field determination of aquifer thermal energy parameters. *Ground Water* 21(1), 22–35.

Rybach, L., Eugster, W.J. (2010). Sustainability aspects of geothermal heat pump operation, with experience from Switzerland. *Geothermics* 39, 365–369.

Sass, J.H., Lachenbruch, A.H., Munroe, R.J. (1971). Thermal conductivity of rocks from measurements on fragments and its application to heat-flow determinations. *Journal of Geophysical Research* 76, 3391–3401.

Sauty, J.P., Gringarten, A.C., Fabris, H., Thiery, D., Menjoz, A. Landel, P.A. (1982b). Sensible energy storage in aquifers. 2. Field experiments and comparison with theoretical results. *Water Resources Research* 18(2), 253–265.

Sauty, J.P., Gringarten, A.C., Menjoz, A., Landel, P.A. (1982a). Sensible energy storage in aquifers. 1. Theoretical study. *Water Resources Research* 18(2), 245–252.

Söll, T. (1985). Vertikal-ebene Modellierung einer Kaltwassereinleitung in das Grundwasser. *Wasserwirtschaft* 75(9), 384–392, Springer, Berlin, SSN: 0043-0978.

Tsang, C.F., Buscheck, T., Doughty, C. (1981). Aquifer thermal energy storage: A numerical simulation of Auburn University field experiments. *Water Resources Research* 17(3), 647–658.

USGS (2012). 3D Finite-difference groundwater flow model MODFLOW. *United States Geological Service*, http://water.usgs.gov/software/lists/groundwater.

Vandenbohede, A., Hermans, T., Nguyen, F., Lebbe, L. (2011). Shallow heat injection and storage experiment: Heat transport simulation and sensitivity analysis. *Journal of Hydrology* 409, 262–272.

Wagner, V., Blum, P. (2012). geomatrix.bw: Teil 2: Prozessmodellierung und Chancenanalyse oberflächennaher Erdwärme in Baden-Württemberg. Report ZO4E28004, Karlsruhe.

Weinszieher, R. (1984). Hydrogeologische und quartärgeologische Untersuchungen im Raum Bad Waldsee-Wolfegg-Bad Wurzach (Lkr. Ravensburg, Oberschwaben). Albert-Ludwigs-Universität, Freiburg.

Xue, Y., Xie, C., Li, Q. (1990). Aquifer thermal energy storage: A numerical simulation of field experiments in China. *Water Resources Research* 26(10), 2365–2375.

Zheng, C., Wang, P.P. (1999). MT3DMS: A modular three-dimensional multi-species transport model for simulation of advection, dispersion and chemical reactions of contaminants in groundwater systems; Documentation and user's guide. U.S. Army Engineer Research and Development Center Contract Report SERDP-99-1, Vicksburg, Mississippi.

Index

Page numbers followed by f and t indicate figures and tables, respectively.

Absorption cycle heat pumps, 10, 11
Advection and three-dimensional
 conduction, in closed systems
 moving point source, 105–106
Advective heat flux, defined, 54
Advective heat transport, 53–54, 70
Aefligen (Switzerland) site, injection of
 cold water, 230
Alabama, USA (Auburn site near
 Mobile)
 aquifer storage and recovery of
 heated water, 229
Altach (Austria), case study, 232–239,
 234t, 235f, 237f, 238f, 239t
Analytical approximations, 101
Analytical models, for TRT evaluation,
 214–218, 215f, 216f, 217f
 error analysis, 218
 Kelvin line source model, 214,
 217–218
 linear regression method, 214–215,
 216
 parameter estimation technique,
 215, 216, 216f
 temporal superposition principle, 216
Analytical solutions
 BHEs, 77
 for flow and heat transport
 problems, 101–158
  for BHE systems, 103
  for closed systems. See Closed systems, analytical solutions for
  literature overview, 101–103
  for open systems. See Open systems, analytical solutions for

overview, 101–104
 thermal sources of, 103, 104f
Anisotropic conditions, defined, 42
Aquifer bottom elevation, defined, 51
Aquifer material, thermal conductivity
 of, 54–55
Aquifers, integral water and energy
 balance equations for, 81–86,
 82f, 83f. See also Unconfined
 aquifer
Aquifers, thermal evolution in
 sustainability and, 202–204
Aquifer thermal energy storage (ATES)
 systems, 164, 165, 167,
 209–210
Aquifer thickness, defined, 51
Aquifer transmissivity, defined, 51
Arithmetic mean model, for thermal
 conductivity, 88
ATES (aquifer thermal energy storage)
 systems, 164, 165, 167,
 209–210
Auburn site near Mobile (Alabama,
 USA), aquifer storage and
 recovery of heated water, 229
Austria, regulatory issues for
 groundwater resources, 29

Back-of-the-envelope calculations, 17
Bad Wurzach (Germany), case study,
 243–248, 244f, 245f, 245t,
 246f, 247f
Belgium (Ghent), heat transport during
 shallow heat injection, 232

BHEs. *See* Borehole heat exchangers
  (BHEs) system
Bonnaud (Jura) site (France), warm
  water storage in a sandy
  gravel aquifer, 229
Borden site (Canada), thermal energy
  storage in unconfined sand
  aquifer, 230
Borehole heat exchangers (BHEs)
  system, 9, 10, 14, 21, 166–167
  analytical solutions for, 77, 103
  Bad Wurzach (Germany), case
    study, 243–248, 244f, 245f,
    245t, 246f, 247f
  boundary condition, 76
  closed-system, 17
  computer simulations for, 167, 186
  concepts for, 75–79, 75f
  heat catchment of, 198–199, 198f
  heat extraction rates for single
    closed system, 18t
  ICS model, 109, 112–113, 113f
  ILS model and, 106–109, 108f,
    109f, 110f–111f
  inadequate backfilling of, 22
  initial condition, 76
  location of, 28
  long-term viability and, 197
  maximum heat flux to, 199
  MILS model, 121–126, 121f, 123f,
    125f, 126f
  one-dimensional finite difference, 219
  SBM in, 219
  sustainability and, 198–200, 198f,
    201f, 202f
  temperature field for, 108, 108f
  timescale for thermal diffusion in, 77
  TRT and, 212
  unit length thermal resistance, 78
  U-tube, 13, 13f, 75–76, 75f, 77, 78
    configuration, 103
  vertical, 13
    two-dimensional finite volume
      model of, 219
Borehole heat load, calculation, 74
Borehole thermal energy storage
  (BTES), 220
Boundary conditions
  of BHEs system, 76
  for flow problems, 49–50, 49f
  for heat transport problem, 72–75,
    73f

Britain, regulatory issues for
  groundwater resources, 29–30
BTES (borehole thermal energy
  storage), 220
Buildings
  heating effect of, 4
  heat loss from, 4
  thermal anomalies and, 3

Canada (Borden site), thermal energy
  storage in unconfined sand
  aquifer, 230
Canton of Zurich, 240
  regulations, 28
Capillary pressure, defined, 47
Capillary zone
  above water table, 41, 41f
  defined, 37
Carbon dioxide (greenhouse gases)
  concentration
  reduction, GSHP systems and, 2–3
Carnot cycle, ideal performance of,
  11–12
Case studies, 229–248
  Aefligen (Switzerland) site, injection
    of cold water, 230
  Alabama, USA (Auburn site near
    Mobile)
    aquifer storage and recovery of
      heated water, 229
  Altach (Austria), 232–239, 234t,
    235f, 237f, 238f, 239t
  Bad Wurzach (Germany), 243–248,
    244f, 245f, 245t, 246f, 247f
  Bonnaud (Jura) site (France), warm
    water storage in a sandy
    gravel aquifer, 229
  Borden site (Canada), thermal
    energy storage in unconfined
    sand aquifer, 230
  Ghent (Belgium), heat transport
    during shallow heat injection,
    232
  Kirchberg (Switzerland) site, natural
    river water infiltration, 230
  Limmat Valley Aquifer Zurich
    (Switzerland), 239–243, 241f,
    242f
  Shanghai (China), seasonal aquifer
    thermal energy storage
    experiments in, 230

Tokyo (Japan), Chiba experimental station in, 232
Torino (Italy), open-loop heat pumps of field study in, 232
Tricks Creek study area (Ontario, Canada), two-dimensional vertical thermal conductivity field, 231
Yucca Mountain site (USA), hydrologic and thermal processes in a large-scale underground heater test, 231
Cauchy boundary condition, for heat transport problem, 73f, 74
Cement–bentonite suspension, 212
Characteristics, defined, 184
China (Shanghai), seasonal aquifer thermal energy storage experiments in, 230
Closed-loop systems, 12–16, 13f, 14f, 15f
Closed systems, analytical solutions for, 105–140
  one-dimensional conduction
    advection and, moving infinite plane source, 132–134, 133f
    infinite plane source, 130, 131
  one-dimensional conductive–advective heat transport
    harmonic temperature boundary condition for. *See* Harmonic temperature boundary condition
  steady-state injection into aquifer with thermally leaky top layer, 134–135
  three-dimensional conduction
    advection and, MFLS, 126–130, 130f, 131f
    advection and, moving point source, 105–106
    FCS model, 118–119
    FLS model, 114–117, 115f, 116f, 117f
  instantaneous point source, 105
  two-dimensional conduction
    advection and, MILS, 119–126, 120f, 121f, 123f, 125f, 126f
    ICS model, 109, 112–113, 113f
    ILS model, 106–109, 108f, 109f, 110f–111f

Closed systems, low hydraulic conductivity formations for, 9, 10
Coded functions (MATLAB scripts)
  analytical solutions
    linear flow, 153
    radial flow, infinite disk source, 156
  FCS model, 119
  FLS model, 116, 117
  ICS model, 112–113
  ILS model, 108
  infinite plane source model, 133–134
  MFLS model, 129
  MILS model, 121, 122, 124, 126
  one-dimensional harmonic thermal conductive/dispersive–advective transport, model for, 138
  recirculation rate, calculation, 151
Coefficient of performance (COP), defined, 11
Complementary error function, in FLS model, 114
Compressibility, of water, 39–40
COMSOL, software package, 167
Conduction, convection, and consolidation (CCC) code, 164
Conductive heat flux, in finite difference method, 177
Constant concentrations, defined, 39
Constant reference pressure, defined, 39
Constant reference temperature, defined, 39
Constant water density, 43–44
Continuous line source, 106–107
Continuous point source
  in FCS, 118
  in FLS model, 114
Continuous ring source model, in FCS, 118
Continuous time random walk (CTRW) model, 67
COP (coefficient of performance), defined, 11
Correction terms, in random walk method, 185
Coupled flow and heat transport, strategy for, 185–186

Coupling thermal transport, with
hydraulic models, 79–80
Courant–Friedrichs–Lewy stability
condition, 180
Courant number, 183
Crank–Nicolson scheme, 179, 181
CTRW (continuous time random walk)
model, 67
Cylindrical source models, 213

Damages, geothermal systems
malfunctioning, 21–22, 23f
Darcy's law, 46
flow in saturated and unsaturated
porous media, 37–44, 38f,
41f, 42f
in volume balance equations, 47
Density effects, stability number for,
40–41
Direct-push methods, 210
Dispersive heat transport, 61–68, 64f,
66f
Distributed optical-fiber temperature
sensing (DTS), 221
Double well system in uniform flow field
analytical solutions, 145–152,
146f–147f
Drainable porosity (specific yield),
defined, 51
Drinking water
quality targets, 24
regulatory issues for groundwater
resources. See Regulatory
issues, for groundwater
resources
DTS (distributed optical-fiber
temperature sensing), 221

Earth Energy Designer (EED)
software, 103
Ecology, groundwater quality and
impact on, 20–21
EED (Earth Energy Designer)
software, 103
Effective thermal conductivity tensor, 62
EGRT (enhanced geothermal response
test), 212
EGS (Engineered Geothermal Systems),
200
Energy balance equation, in finite
difference method, 176

Energy demand, 16–19, 17t, 18t
seasonal, 16f
Energy fluxes, toward soil surface, 8, 8f
Energy production, 16–19, 16f, 18t
by GSHPs, 17t
Energy utilization factor, defined, 86
Engineered Geothermal Systems (EGS),
200
Enhanced geothermal response test
(EGRT), 212
Enhanced thermal response tests, 212
Equivalent decay coefficient, in solute
transport code, 171
Equivalent porosity, in solute transport
code, 170
Equivalent solute mass production, in
solute transport code, 170
Exponential integral, definition, 107

FCS (finite cylindrical source), in
closed systems
three-dimensional conduction,
118–119
FEFLOW code, 219, 220
FEFLOW software, 165, 166, 167, 186
Field methods, 209–224
hydrogeological, 209–210
overview, 209
TRTs. See Thermal response tests
(TRTs)
TTTs. See Thermal tracer tests
(TTTs)
Finite cylindrical source (FCS), in
closed systems
three-dimensional conduction,
118–119
Finite difference method, for heat
transport
principles, in multidimensional
numerical solutions, 166, 175,
176–181, 177f
Finite element method, for heat transport
principles, in multidimensional
numerical solutions, 164, 175,
181–183, 181f
Finite line source (FLS), 103, 104f
in closed systems, three-dimensional
conduction, 114–117, 115f,
116f, 117f
advection and, MFLS, 126–130,
130f, 131f

Finite volume method, for heat transport principles, in multidimensional numerical solutions, 166, 175, 183

First- or second-order predictor–corrector schemes for coupled equations solution, 186

Flow equation, hydraulic processes in porous media, 48

FLS (finite line source), 103, 104f
in closed systems, three-dimensional conduction, 114–117, 115f, 116f, 117f
advection and, MFLS, 126–130, 130f, 131f

FLUENT software, 167

Fokker–Planck terms, in random walk method, 185

FORTRAN-based code, 219

Fourier's law, defined, 54

FRACTure, for forced water flow in fractured rocks, 165

FRACTure code, for TRT analysis, 219

France (Bonnaud (Jura) site), warm water storage in a sandy gravel aquifer, 229

Freezing and thawing, heat transport equation, 68–69

Frozen soil conditions, 60

Fundamentals, of heat transport in groundwater systems, 37–93
heat transport processes in subsurface. See Heat transport processes in subsurface, water flow and thermal property values, 87–93, 87t, 88t, 89t–91t, 92t, 93f
water flow theory in subsurface. See Water flow, heat transport processes in subsurface and

Galerkin finite element model, 164–165

GenOpt software, 219

Geometric mean model, for thermal conductivity, 88

Geotechnical constructions, heat exchangers in, 14, 15

Geotechnical issues, 21–22, 23f

Geothermal ground source heating pump (GSHP) system

Bad Wurzach (Germany), case study, 243–248, 244f, 245f, 245t, 246f, 247f

Germany
Bad Wurzach, case study, 243–248, 244f, 245f, 245t, 246f, 247f
regulatory issues for groundwater resources, 30–31

Ghent (Belgium), heat transport during shallow heat injection, 232

GLHEPRO software, 103

Greenhouse gases (carbon dioxide) concentration reduction, GSHP systems and, 2–3

Grid Peclet number, 183

Ground heat flux, defined, 8

Ground source heat pump (GSHP) systems, 1, 2f, 3, 12, 21–22, 209
TRTs for planning and design of. See Thermal response tests (TRTs)

Ground surface temperature (GST), 4–9, 5f, 6f, 7f, 8f, 9f

Groundwater heat pump (GWHP) systems, 1, 2f

Groundwater quality and ecology, impact on, 20–21

Groundwater Regulations, 30

Groundwater systems, thermal use. See Thermal use of shallow underground systems

GSHP (ground source heat pump) systems, 1, 2f, 3, 12, 21–22, 209
TRTs for planning and design of. See Thermal response tests (TRTs)

GST (ground surface temperature), 4–9, 5f, 6f, 7f, 8f, 9f

GWHP (groundwater heat pump) systems, 1, 2f

Hantush approximation, 156

Harmonic mean model, for thermal conductivity, 88

Harmonic temperature boundary condition for one-dimensional conductive–advective heat transport, 135–140

horizontal conductive/dispersive-
  advective transport, 136–138,
  137f, 138f
horizontal layer embedded in
  conductive bottom and top
  layer, 138–140
vertical conductive heat
  transport, 135–136
Heat capacity
  thermal conductivity values and,
    87–93, 87t, 88t, 89t–91t, 92t,
    93f
  thermal processes in porous media,
    53–54
"Heat catchment," of BHEs, 198–199,
  198f
Heat conduction, thermal processes in
  porous media. *See* Thermal
  conductivity(ies)
Heat exchangers, in geotechnical
  constructions, 14, 15
HEATFLOW, finite element numerical
  model, 231
HEATFLOW-SMOKER code, 186
Heat pumps, 10–12
  absorption cycle, 10, 11
  COP, 11
  ideal performance of Carnot cycle,
    11–12
  PER, 12
  SPF, 12
  vapor compression, 10, 11f
  working fluid of, regulations, 24
Heat sink, 3
Heat storage, thermal processes in
  porous media, 53–54
Heat transport equations, coupled flow
  and
  strategy for, 185–186
Heat transport processes in subsurface,
  water flow and, 37–86
  hydraulic processes in porous
    media, 37–52
    flow equation, 48
    flow in saturated and unsaturated
      porous media, Darcy's law,
      37–44, 38f, 41f, 42f
    initial and boundary conditions,
      49–50, 49f
    two-dimensional flow models for
      saturated regional water flow,
      50–52, 51f, 52f

volume balance, 47–48
water mass balance, 44–47, 45f
integral water and energy balance
  equations for aquifers, 81–86,
  82f, 83f
overview, 37
thermal processes in porous media,
  52–81
  BHEs, concepts for, 75–79, 75f
  coupling thermal transport with
    hydraulic models, 79–80
  dispersive and macrodispersive,
    61–68, 64f, 66f
  heat conduction. *See* Thermal
    conductivity(ies)
  heat storage, heat capacity, and
    advective heat transport,
    53–54
  heat transport equation, 68–72
  initial and boundary conditions,
    72–75, 73f
  two-dimensional heat transport
    models, 80–81, 80f
Horizontal groundwater flow, TRT
  and, 218
Horizontal layer embedded in
  conductive bottom and top
  layer, harmonic temperature
  boundary condition for,
  138–140
HST3D code, 186
Hydraulic conductivity
  in anisotropic conditions, 42
  in layered porous media, 42
  temperature dependence of water
    viscosity on, 38–39
Hydraulic contacts, boreholes and,
  10
Hydraulic inflow rates, for unconfined
  aquifer, 81–82, 82f
Hydraulic models, coupling thermal
  transport with, 79–80
Hydraulic outflow rates, for unconfined
  aquifer, 81–82, 82f
Hydraulic processes, in porous media,
  37–52
  flow equation, 48
  flow in saturated and unsaturated
    porous media, Darcy's law,
    37–44, 38f, 41f, 42f
  initial and boundary conditions,
    49–50, 49f

two-dimensional flow models for
    saturated regional water flow,
    50–52, 51f, 52f
volume balance, 47–48
water mass balance, 44–47, 45f
Hydraulic tomography method, 210
Hydrogeological conditions, 9, 10
Hydrogeological field methods,
    209–210
HydroGeoSphere code, 165–166, 187,
    219
Hydrographic Service, 234
Hydrological conditions, 9, 10

ICS (infinite cylindrical source), in
    closed systems
    two-dimensional conduction, 109,
        112–113, 113f
ILS (infinite line source), 103, 104f
    in closed systems, two-dimensional
        conduction, 106–109, 108f,
        109f, 110f–111f
    advection and, MILS, 119–126,
        120f, 121f, 123f, 125f, 126f
Impermeable boundary, 49f, 50
Inadequate backfilling, cause of
    damages, 22
Induction of thermal fluxes, 3
Infinite cylindrical source (ICS), in
    closed systems
    two-dimensional conduction, 109,
        112–113, 113f
Infinite disk source, radial flow
    analytical solutions, 155–156, 157f
Infinite line source (ILS), 103, 104f
    in closed systems, two-dimensional
        conduction, 106–109, 108f,
        109f, 110f–111f
    advection and, MILS, 119–126,
        120f, 121f, 123f, 125f, 126f
Infinite plane source
    in closed systems, one-dimensional
        conduction, 130, 131
    advection and, moving infinite
        plane source, 132–134, 133f
Initial conditions
    of BHEs system, 76
    for flow problems, 49–50, 49f
    for heat transport problem, 72–75,
        73f
Instantaneous line source, 106

Instantaneous point source, in closed
    systems
    three-dimensional conduction, 105
Integral water and energy balance
    equations, for aquifers,
    81–86, 82f, 83f
Integrated compartment method, 166
International Governmental
    Commission of Alpine Rhine
    region (IRKA), 236
IRKA (International Governmental
    Commission of Alpine Rhine
    region), 236
Italy (Torino), open-loop heat pumps
    of field study in, 232

Japan (Tokyo), Chiba experimental
    station in, 232
Johansen's model, thermal conductivity
    and, 58, 59

Kelvin line source theory, 213
    in TRT evaluation, 214, 217–218
Kersten's number, thermal conductivity
    and, 59–60
Kirchberg (Switzerland) site, natural
    river water infiltration, 230

Land subsidence and uplift, cause of
    damages, 22, 23f
Land surface temperature (LST),
    satellite-derived, 7
Lawrence Berkeley Laboratory, 164
Leakage coefficient, 50, 52
Limmat Valley Aquifer Zurich
    (Switzerland), case study,
    239–243, 241f, 242f
Linear flow, analytical solutions for
    in open systems, 152–153, 152f, 154f
Linear regression method, for TRTs,
    214–215, 216
Local meteorological conditions,
    importance, 4–10
    hydrological and hydrogeological
        conditions, 9, 10
    thermal regime, 4–9, 5f, 6f, 7f, 8f, 9f
Longitudinal thermal
    macrodispersivity, 64–66, 66f
Long-term injection-storage
    experiments, 220

Long-term operability, 197–207
  further criteria of sustainability,
    204–207, 205t, 206t
  overview, 197
  systems in low permeable media,
    197–202, 198f, 201f, 202f
  thermal evolution in aquifers,
    202–204
Low-enthalpy energy (shallow
    geothermal energy), 1
Low permeable media, systems in
  sustainability and, 197–202, 198f,
    201f, 202f
LST (land surface temperature),
    satellite-derived, 7

Macrodispersive heat transport,
    61–68, 64f, 66f
MATLAB scripts
  analytical solutions
    linear flow, 153
    radial flow, infinite disk source,
      156
  FCS model, 119
  FLS model, 116, 117
  ICS model, 112–113
  ILS model, 108, 109
  infinite plane source model,
    133–134
  MFLS model, 129
  MILS model, 121, 122, 124, 126
  one-dimensional harmonic thermal
    conductive/dispersive–
    advective transport, model
    for, 138
  recirculation rate, calculation, 151
Maximum heat flux, to BHEs, 199
Mechanical dispersion effects, in porous
    media, 61–68, 64f, 66f
Method of characteristics, for heat
    transport
  principles, in multidimensional
    numerical solutions, 183–184
METRA, advective–conductive heat
    transport models with, 165
MFLS (moving finite line source), in
    closed systems
  three-dimensional conduction and
    advection, 126–130, 130f, 131f
MILS (moving infinite line source), in
    closed systems

two-dimensional conduction and
    advection, 119–126, 120f,
    121f, 123f, 125f, 126f
MODFLOW96 code, 236
MODFLOW model, 187
Moving finite line source (MFLS), in
    closed systems
  three-dimensional conduction and
    advection, 126–130, 130f,
    131f
Moving finite line source model, 213
Moving infinite line source (MILS), in
    closed systems
  two-dimensional conduction and
    advection, 119–126, 120f,
    121f, 123f, 125f, 126f
Moving infinite plane source, in closed
    systems
  one-dimensional conduction and
    advection, 132–134, 133f
Moving point source, in closed systems
  three-dimensional conduction and
    advection, 105–106
MT3DMS code, 236
  in heat transport in closed
    geothermal systems, 165,
    187
Multidimensional numerical solutions,
    175–185
  heat transport, principles
    finite difference method for, 166,
      175, 176–181, 177f
    finite element method for, 164,
      175, 181–183, 181f
    finite volume method for, 166,
      175, 183
    method of characteristics for,
      183–184
    random walk method for, 165,
      175, 184–185
MULTIFLOW code, 165
Multiple well systems, steady-state
    flow in
  analytical solutions for, open
    systems, 142–152, 143f, 144f
  double well system in uniform
    flow field, 145–152, 146f–147f

Natural background groundwater flow,
    in open systems
  analytical solutions, 156, 158

Natural conditions, defined, 26
Net radiation, defined, 8
Neumann boundary condition
    for flow problems, 50
    for heat transport problem, 72, 73f
Numerical codes for thermal transport
        modeling in ground water,
        186–187, 188t–189t
    HEATFLOW-SMOKER code, 186
    HST3D code, 186
    HydroGeoSphere code, 165–166, 187
    MODFLOW model, 187
    MT3DMS, 165, 187
    MULTIFLOW code, 165
    SEAWAT code, 165, 187
    SHEMAT code, 187
    SPRING code, 166, 187
    SUTRA code, 187
    TOUGH2 code, 166, 187
    VS2DI code, 187
Numerical models, for TRT evaluation,
        218–220, 220f
Numerical solutions, 163–189
    codes for thermal transport
        modeling in ground water,
        186–187, 188t–189t
    HEATFLOW-SMOKER code,
        186
    HST3D code, 186
    HydroGeoSphere code, 165–166,
        187
    MODFLOW model, 187
    MT3DMS, 165, 187
    MULTIFLOW code, 165
    SEAWAT code, 165, 187
    SHEMAT code, 187
    SPRING code, 166, 187
    SUTRA code, 187
    TOUGH2 code, 166, 187
    VS2DI code, 187
    development of, 163
    multidimensional, for heat
        transport, 175–185
    finite difference method,
        principles, 166, 175, 176–181,
        177f
    finite element method, principles,
        164, 175, 181–183, 181f
    finite volume method, principles,
        166, 175, 183
    method of characteristics,
        principles, 183–184

    random walk method, principles,
        165, 175, 184–185
    overview, 163–167
    strategy for coupled flow and heat
        transport, 185–186
    two-dimensional horizontal, 167–175
    solute transport models, analogy
        with, 170–171
    steady-state open system in
        rectangular aquifer, analysis,
        171–175, 171f, 172f, 173f,
        174f, 175f

One-dimensional conduction, in closed
        systems
    advection and, moving infinite
        plane source, 132–134, 133f
    infinite plane source, 130, 131
One-dimensional conductive–advective
        heat transport
    harmonic temperature boundary
        condition for. See Harmonic
        temperature boundary
        condition
One-dimensional finite difference BHE
        model, 219
One-dimensional horizontal conductive/
        dispersive-advective transport
    harmonic temperature boundary
        condition for, 136–138, 137f,
        138f
One-dimensional vertical conductive
        heat transport
    harmonic temperature boundary
        condition for, 135–136
Ontario, Canada (Tricks Creek study
        area), two-dimensional vertical
        thermal conductivity field, 231
Open-loop systems, 12–16, 13f, 14f, 15f
Open systems, analytical solutions for,
        140–158, 141t
    linear flow, 152–153, 152f, 154f
    natural background groundwater
        flow, 156, 158
    radial flow, infinite disk source,
        155–156, 157f
    steady-state flow in multiple well
        systems, 142–152, 143f, 144f
    double well system in uniform
        flow field, 145–152, 146f–147f
Open-well thermal dilution tests, 221

Parameter estimation technique, for TRTs, 215, 216, 216f
Particle tracking, in random walk method, 184
PER (primary energy ratio), defined, 12
Picard iteration scheme (point iteration), 185
PMWIN, software package, 236
Pollutants, water quality by, 21, 24
Porous media
  hydraulic processes, flow in saturated and unsaturated, 37–44, 38f, 41f, 42f. *See also* Hydraulic processes, in porous media
  thermal processes in. *See* Thermal processes, in porous media
PPTs (push–pull tests), 221
Predictor–corrector schemes, first- or second-order
  for coupled equations solution, 186
Primary energy ratio (PER), defined, 12
Pumped groundwater as heat source, 3
Push–pull tests (PPTs), 221

QS/QA (quality standards and quality assurances)
  future damages and, 22
Quality standards and quality assurances (QS/QA)
  future damages and, 22

Radial flow, infinite disk source analytical solutions, 155–156, 157f
Random walk method, for heat transport
  principles, in multidimensional numerical solutions, 165, 175, 184–185
ReacTrans, 221
Recirculation rate, between wells, 149–151
Rectangular aquifer, Steady-state open system in
  analysis, two-dimensional horizontal numerical solutions, 171–175, 171f
  illustrative example, 172f, 173f
  scaled solution, 173–175, 174f, 175f

Regional water flow, two-dimensional flow models for, 50–52, 51f, 52f
Regulatory issues, for groundwater resources, 24–31
  Austrian regulation, 29
  British regulation, 29–30
  for drinking water supply, 24
  exploitation of renewable resource, 24
  German regulation, 30–31
  heat pumps, 24
  international legal status, 25
  recommendations and, 25
  for soil and ecosystem protection, 24
  Swiss regulation, 25–28, 27f
Renewable resources, exploitation of, 24. *See also* Regulatory issues, for groundwater resources
Renormalization (self-consistent) approximation, 55
Richards' equation, defined, 47

Satellite-derived LST, 7
Saturated porous media, flow in
  Darcy's law, 37–44, 38f, 41f, 42f
Saturated regional water flow, two-dimensional flow models for, 50–52, 51f, 52f
SBM (superposition borehole model), 219
Seasonal operation of technical installations
  underground resources management, 19–20
Seasonal performance factor (SPF), defined, 12
Seasonal soil surface temperature, 8
SEAWAT code, 165, 187
Self-consistent (renormalization) approximation, 55
Semipermeable boundary, flux condition for, 49f, 50
Sewer systems, thermal anomalies and, 3
Shallow geothermal energy (low-enthalpy energy), 1
Shallow groundwater systems, thermal use. *See* Thermal use of shallow underground systems

Shallow regional aquifers
    heat transport equation for, 80
    steady-state heat transport in, 80
Shallow soils, 21
Shanghai (China), seasonal aquifer
        thermal energy storage
        experiments in, 230
SHEMAT code, 187
SIA (Swiss Society of Engineers and
        Architects), 26–28
Soil and ecosystem protection,
        regulations, 24. *See also*
        Regulatory issues, for
        groundwater resources
Soil surface temperature, 7–8
    heat balance at, 8, 8f
    seasonal, 8
Solid materials, thermal conductivity
        of, 54
Solute transport equation with first-
        order decay, codes for, 171
Solute transport models, analogy with
    in two-dimensional horizontal
        numerical solutions, 170–171
Specific advective heat flux, defined, 54
Specific heat capacity, thermal
        conductivity values and, 87–93,
        87t, 88t, 89t–91t, 92t, 93f
Specific storativity, concept of, 45, 48
Specific yield (drainable porosity),
        defined, 51
SPF (seasonal performance factor),
        defined, 12
SPRING code, 166, 187
Stability number, defined, 40
Stable thermal yield, 3
Standing column wells, 15
State of Vorarlberg, 233, 234
Steady-state flow in multiple well systems
    analytical solutions for, open
        systems, 142–152, 143f, 144f
    double well system in uniform
        flow field, 145–152, 146f–147f
Steady-state injection, in closed
        systems
    into aquifer with thermally leaky
        top layer, 134–135
Steady-state open system in rectangular
        aquifer, analysis
    two-dimensional horizontal
        numerical solutions, 171–175,
        171f

illustrative example, 172f, 173f
    scaled solution for open system
        in rectangular aquifer,
        173–175, 174f, 175f
Steady-state point source solution, in
        FLS model, 114
Steady-state scaled solution, for open
        system in rectangular aquifer,
        173–175, 174f, 175f
Superposition borehole model (SBM),
        219
Surface heat fluxes, in finite volume
        method, 183
Sustainability, of thermal use of
        the shallow underground,
        197–207
    BHEs and, 198–200, 198f, 201f, 202f
    further criteria, 204–207, 205t, 206t
    overview, 197
    systems in low permeable media,
        197–202, 198f, 201f, 202f
    thermal evolution in aquifers,
        202–204
SUTRA code, 187
Swiss Society of Engineers and
        Architects (SIA), 26–28
Switzerland
    Aefligen, injection of cold water, 230
    Kirchberg, natural river water
        infiltration, 230
    Limmat Valley Aquifer Zurich, case
        study, 239–243, 241f, 242f
    regulatory issues for groundwater
        resources, 25–28, 27f
    Zurich-Affoltern
        ground temperature profiles at
            meteorological station, 5–6, 6f
        monthly average temperature of
            air and soil at meteorological
            station, 6, 7f

TAZ (thermally affected zone), 244
Technical guideline, for quality
        assurance for GSHP systems,
        22
Technical systems, 10–16
    closed- and open-loop systems,
        12–16, 13f, 14f, 15f
    heat pumps. *See* Heat pumps
TED (TRT device), 210, 211f, 212
Temperature anomalies, 3, 199

Temperature range, defined, 16
Temporal superposition principle, in
    TRT evaluation, 216
Thawing, freezing and
    heat transport equation, 68–69
Thermal anomalies, 3, 199
Thermal borehole resistance,
    determination, 74
Thermal conductivity(ies), 54–61
    of aquifer materials, 54–55
    de Vries method for, 56–57
    for dry conditions, 59
    effective, 55–56
    heat conduction equation, 71
    Johansen's model, 58, 59
    Kersten's number and, 59–60
    Kunii and Smith, equations of, 56
    porosity and saturation degree on, 60
    of saturated and unsaturated porous
        media, 55
    self-consistent (renormalization)
        approximation and, 55
    soils, 56–57
    solid materials, 54
    thermal properties and densities of
        basic soil constituents, 60t
    values, heat capacity and, 87–93,
        87t, 88t, 89t–91t, 92t, 93f
    water, 54
    weighted arithmetic average, 55
    by weighted geometric average,
        58–59
    for wet material with water
        saturation, 58–59
Thermal diffusion coefficient, of
    porous medium, 69
Thermal evolution in aquifers,
    sustainability and, 202–204
Thermal exploitation, of resource, 3
Thermal fluxes, induction of, 3
Thermal inflow rates, for unconfined
    aquifer, 83–84, 83f
Thermally affected zone (TAZ), 244
Thermally insulating boundary, for
    heat transport problem, 73,
    73f
Thermally leaky top layer, in closed
    systems
    steady-state injection into aquifer
        with, 134–135
Thermal mechanical dispersion effect,
    61–68, 64f, 66f

Thermal monitoring, of groundwater
    resources. See Regulatory
    issues, for groundwater
    resources
Thermal outflow rates, for unconfined
    aquifer, 83–84, 83f
Thermal overexploitation, 3
Thermal Peclet number, 63, 71
    MILS model and, 120
Thermal processes, in porous media,
    52–81
    BHEs, concepts for, 75–79, 75f
    coupling thermal transport with
        hydraulic models, 79–80
    dispersive and macrodispersive heat
        transport, 61–68, 64f, 66f
    heat conduction. See Thermal
        conductivity(ies)
    heat storage, heat capacity, and
        advective heat transport,
        53–54
    heat transport equation, 68–72
    initial and boundary conditions,
        72–75, 73f
    two-dimensional heat transport
        models, 80–81, 80f
Thermal property values, heat capacity
    and thermal conductivity
    values, 87–93, 87t, 88t,
    89t–91t, 92t, 93f
Thermal regime, local meteorological
    conditions and, 4–9, 5f, 6f,
    7f, 8f, 9f
Thermal response tests (TRTs), 209,
    210–220
    development, 210–212, 211f
    evaluation, 213–220
        analytical models. See Analytical
            models, for TRT evaluation
        numerical models, 218–220, 220f
    setup and application, 212–213,
        213f
Thermal retardation factor, 70
Thermal tracer tests (TTTs), 209,
    220–224, 222t–223t
Thermal use of shallow underground
    systems
    analytical solutions. See Analytical
        solutions
    case studies. See Case studies
    challenges related to design and
        management, 31

energy demand and energy production, 16–19, 16f, 17t, 18t
field methods. *See* Field methods
fundamentals. *See* Fundamentals, of heat transport in groundwater systems
geotechnical issues, 21–22, 23f
groundwater quality and ecology, impact on, 20–21
local conditions, importance. *See* Local meteorological conditions, importance
long-term operability. *See* Long-term operability
motivation for, 2–4
numerical solutions. *See* Numerical solutions
overview, 1–32
regulatory issues. *See* Regulatory issues, for groundwater resources
sustainability. *See* Sustainability
technical systems. *See* Technical systems
underground resources, management. *See* Underground resources, management
Three-dimensional conduction, in closed systems
advection and
MFLS, 126–130, 130f, 131f
moving point source, 105–106
FCS model, 118–119
FLS model, 114–117, 115f, 116f, 117f
instantaneous point source, 105
Three-dimensional finite element code, 219
Tokyo (Japan), Chiba experimental station in, 232
Torino (Italy), open-loop heat pumps of field study in, 232
TOUGH2 code, 166, 187
Transmission boundary condition for heat transport problem, 74
Transversal thermal macrodispersivity, 64–65, 66–67
Trial solution, defined, 181
Tricks Creek study area (Ontario, Canada)
two-dimensional vertical thermal conductivity field, 231

TRNSBM software, 219
TRNSYS software, 219
TRT device (TED), 210, 211f, 212
TRTs. *See* Thermal response tests (TRTs)
TTTs (thermal tracer tests), 209
Two-dimensional conduction, in closed systems
advection and, moving ILS, 119–126, 120f, 121f, 123f, 125f, 126f
ICS model, 109, 112–113, 113f
ILS model, 106–109, 108f, 109f, 110f–111f
Two-dimensional finite volume model, of vertical BHE, 219
Two-dimensional flow models
for saturated regional water flow, 50–52, 51f, 52f
Two-dimensional heat transport models, 80–81, 80f
Two-dimensional horizontal numerical solutions, 167–175
solute transport models, analogy with, 170–171
steady-state open system in rectangular aquifer, analysis, 171–175, 171f
illustrative example, 172f, 173f
scaled solution for open system in rectangular aquifer, 173–175, 174f, 175f

UFM (Upper Freshwater Molasse), 243
Unconfined aquifer
hydraulic inflow and outflow rates for, 81–82, 82f
shallow aquifer for thermal use, potential estimation, 85–86
thermal inflow and outflow rates for, 83–84, 83f
Underground resources, management, 19–20
seasonal operation of technical installations, 19–20
water supply and thermal use, 20
Underground systems, thermal use. *See* Thermal use of shallow underground systems
Uniform flow field, double well system in analytical solution, 145–152, 146f–147f

United States
    Alabama (Auburn site near Mobile),
        aquifer storage and recovery
        of heated water, 229
    Yucca Mountain site, hydrologic
        and thermal processes in
        a large-scale underground
        heater test, 231
Unsaturated porous media, flow in
    Darcy's law, 37–44, 38f, 41f, 42f
Upper Freshwater Molasse (UFM), 243
Urban environments, mean annual
        ground temperatures, 7
Urban heat island effect, 3–4
Urbanization on shallow groundwater,
        4
Utilization of underground, 3
U-tube configuration, of BHEs, 13,
        13f, 103
    single, 75–76, 75f, 77, 78

Van-'t-Hoff rule, 21
Vapor compression heat pumps, 10, 11f
Vertical double wells, 15–16
Vertical geothermal heat probes
    installation of, 27
    steps for design of, 27–28
Vertical groundwater flow, TRT and,
        218
Vertical heat balance equation, for
        unconfined aquifers, 169
Vertical Richards equation, 169
Volumetric heat capacity, 53, 54, 87,
        88t, 89t–91t
    thermal conductivity of soil and,
        87t
    thermal conductivity values and,
        87–93, 87t, 88t, 89t–91t, 92t,
        93f
Volumetric water content, defined, 43
VS2DI code, 187

Water density, 40, 41
    flow in saturated porous media with
        constant, 42
    temperature dependence of, 38–39,
        38f
Water flow, heat transport processes in
        subsurface and, 37–86
    hydraulic processes in porous
        media, 37–52

flow equation, 48
flow in saturated and unsaturated
        porous media, Darcy's law,
        37–44, 38f, 41f, 42f
initial and boundary conditions,
        49–50, 49f
two-dimensional flow models for
        saturated regional, 50–52,
        51f, 52f
volume balance, 47–48
water mass balance, 44–47, 45f
integral water and energy balance
        equations for aquifers, 81–86,
        82f, 83f
overview, 37
thermal processes in porous media,
        52–81
    BHEs, concepts for, 75–79, 75f
    coupling thermal transport with
        hydraulic models, 79–80
    dispersive and macrodispersive
        heat transport, 61–68, 64f, 66f
    heat conduction. See Thermal
        conductivity(ies)
    heat storage, heat capacity, and
        advective heat transport,
        53–54
    heat transport equation, 68–72
    initial and boundary conditions,
        72–75, 73f
    two-dimensional heat transport
        models, 80–81, 80f
Water mass balance, hydraulic
        processes in porous media,
        44–47, 45f
Water pressure–dependent piezometric
        head, defined, 46
Water quality
    and ecology, impact on, 20–21
    by pollutants, 21, 24
Water Resources Act, 30
Water retention curve, defined, 47
Water saturation, defined, 44
Water supply, underground resources
        management
    Austria, regulatory issues, 29
    for domestic and industrial use, 19
    Switzerland, regulatory issues,
        25–28, 27f
    thermal use and, 20
Water table, capillary zone and, 41, 41f
Water temperature, 20–21

Water viscosity, 40, 41
  temperature dependence of, 38–39,
      38f
Water volume balance, hydraulic
      processes in porous media,
      47–48
Weighted arithmetic average
  defined, 53
  of thermal conductivity, 55
Weighted geometric average, thermal
      conductivity by, 58–59
Weighted harmonic average, 55
Well-doublet scheme, 15, 15f
Well systems, steady-state flow in
      multiple
  analytical solutions for, open
      systems, 142–152, 143f, 144f

double well system in uniform
    flow field, 145–152, 146f–147f
Working fluid of heat pumps, 24

Yucca Mountain site (USA)
  hydrologic and thermal processes
      in a large-scale underground
      heater test, 231

Zurich-Affoltern (Switzerland)
  ground temperature profiles at
      meteorological station, 5–6,
      6f
  monthly average temperature of air
      and soil at meteorological
      station, 6, 7f

Printed and bound by CPI Group (UK) Ltd, Croydon, CR0 4YY

18/10/2024

01776262-0007